Optical Waveguide Technology and Applications

Edited by Ki Young Kim

Published in London, United Kingdom

Optical Waveguide Technology and Applications
http://dx.doi.org/10.5772/intechopen.114644
Edited by Ki Young Kim

Contributors
Ahmed Aljaroudi, Alireda Aljaroudi, António Teixeira, Carla Rodrigues, Farjana Rahman, Fatemeh Haddadan, Inês Venâncio, Janis Braunfelds, Joana Tátá, Jorge Martínez Garfias, João Santos, Jurgis Porins, Mohammad Javad Maleki, Mohammad Soroosh, Nauris Silkans, Paulo Duarte, Rebeca Diaz Pardo, Sandis Spolitis, Shiva Hayati Raad, Ugis Senkans, Veer Chandra, Vjaceslavs Bobrovs

First published in London, United Kingdom, 2024 by IntechOpen
IntechOpen is the global imprint of INTECHOPEN LIMITED, registered in England and Wales, registration number: 11086078, 167-169 Great Portland Street, London, W1W 5PF, United Kingdom

British Library Cataloguing-in-Publication Data
A catalogue record for this book is available from the British Library

Additional hard and PDF copies can be obtained from orders@intechopen.com

Optical Waveguide Technology and Applications
Edited by Ki Young Kim
p. cm.
Print ISBN 978-0-85014-426-0
Online ISBN 978-0-85014-427-7
eBook (PDF) ISBN 978-0-85014-428-4

For EU product safety concerns:
IN TECH d.o.o., Prolaz Marije Krucifikse Kozulić 3, 51000 Rijeka, Croatia,
info@intechopen.com or visit our website at intechopen.com.

Meet the editor

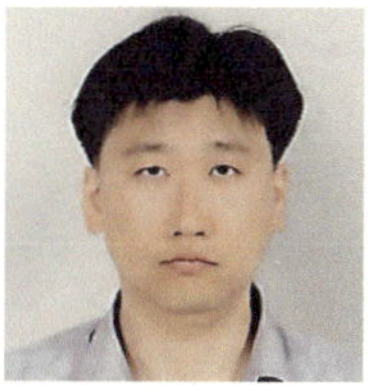

Ki Young Kim, Ph.D., MBA, has been with Samsung Electronics since 2010, where he is currently a Principal Engineer and Corporate Professor. Prior to joining Samsung, he was an associate research professor at Kyungpook National University, Daegu, Korea; an assistant research professor at National Cheng Kung University, Tainan, Taiwan; a senior engineer at Hyundai Motor Company, Hwasung, Korea; a postdoctoral research fellow at Northwestern University, Evanston, USA; and an adjunct researcher at Pohang Accelerator Laboratory, Pohang, Korea. He has published one book chapter, 5 edited books, and more than 120 research journal/conference articles. Dr. Kim has more than 100 US patents registered in his name.

Contents

Preface

Optical waveguides are basic structural elements for the efficient transmission of light over various distances, whether short or long. Traditionally, the primary focus and interest in optical waveguides, particularly optical fibers, has been on long-distance transmission with minimal losses and distortions. In practical terms, significant advancements in research and development have enabled the realization of modern optical and photonic devices, systems, and even services, addressing both fundamental scientific principles and practical commercial applications.

This book presents contemporary perspectives on optical waveguides, focusing on both the technology itself and its applications. It includes seven chapters organized into two sections: "Optical Sensing" and "Waveguiding Media." These sections explore a range of topics, including fiber optical sensing techniques and their underlying theory and applications, integrated optical waveguide structures and performances, waveguides in magnetism and spintronics, graphene-based surface plasmonics, optical waveguides in quantum computations, and optical waveguide fabrication processes.

The book reviews current research trends and developments in optical waveguide technology and. The insights and technologies discussed here may also be applicable or offer valuable perspectives to other areas within the field of optical waveguides. I hope that this collection inspires readers to initiate or further their research, expand potential applications, and explore other aspects of this fascinating area.

I would like to express my sincere gratitude to all the authors for their outstanding contributions to this book. I would also like to thank Publishing Process Manager Ms. Maja Bozicevic, for her kind and dedicated support.

Ki Young Kim
Samsung Electronics,
Yongin, Korea

Section 1

Optical Sensing

Chapter 1

Development of Fiber Bragg Gratings for the Optical Sensor Solutions in Structural Health Monitoring

Janis Braunfelds, Ugis Senkans, Farjana Rahman, Nauris Silkans, Sandis Spolitis, Jurgis Porins and Vjaceslavs Bobrovs

Abstract

Fiber optical sensors (FOS) have been widely used to ensure physical parameter monitoring such as strain, temperature, vibration, etc. Fiber Bragg grating (FBG) sensors are of interest mainly as they offer relatively easy integration, multiplexing capabilities, and other advantages. A topical area of investigation for FOS integration is structural health monitoring (SHM) and perimeter-security monitoring. Different objects and infrastructures have certain needs, such as long-distance (40+ km) placement and monitoring solutions, lack of active power supply availability on the site, etc. It is vital to study and develop specific FBG profiles to ensure optimal operation of FBG in security, perimeter, and SHM solutions. In this research, we have evaluated the areas and categories in which such FOS and FBGs, in particular, have been used regarding SHM and perimeter monitoring solutions. Authors have also studied, simulated, and experimentally tested various FBGs that can be implemented for the perimeter—security applications and in SHM solutions.

Keywords: fiber optical sensors (FOS), fiber Bragg grating (FBG), structural health monitoring (SHM), sensor network, FBG profiles, apodization

1. Introduction

The number of sites and territories in Latvia and worldwide that need different kinds of real-time health monitoring (roads, barrages, bridges, pipelines, dams, railways, buildings, and other structures) is constantly growing. Over the last decade, monitoring of various processes and sites and sensor-based safety control of the public infrastructure has extensively increased [1, 2]. Since monitoring often should be conducted simultaneously at long distances across many locations, in corrosive environments under electromagnetic interference, or in environments without power supply, FBG optical sensors can be applied in such types of monitoring [3–5].

Nowadays, the infrastructure of optical networks has not been exploited and utilized to its full capacity. The authors see high potential in extending the infrastructure applications using optical fiber sensors. Considering that the optical network infrastructure is located at/near roads, bridges, tunnels, railways, dams, etc., it can be used as a transmission medium for monitoring those infrastructure objects. "Dark" optical fiber lines can be applied in such monitoring solutions or working lines can be used, utilizing their free frequency ranges. In order to promote further development of such solutions, it is essential to develop high-value optical FBG sensor networks for long-distance monitoring (40+ km) solutions, to provide interoperability of communication and sensor solutions, and to ensure integration of high-quality sensors in various infrastructure objects for monitoring purposes.

To observe and more comprehensively evaluate the topic's importance and objective topicality, we have analyzed FBG optical sensor-based number of published articles that have been indexed in the SCOPUS database (see **Figure 1**). For our search, we used article abstracts, their titles, and keywords for finding references regarding "fiber optical sensors," "fiber Bragg grating sensors," and "distributed fiber optical sensors." As observable from **Figure 1**, all of our mentioned "search parameters" show clear increased interest, reaching more than 3000 articles per year consisting of keywords related to "fiber optical sensors," more than 1000 articles per year for "fiber Bragg grating sensors" and more than 500 – for "distributed fiber optical sensors." The available data shows a stable increase in interest in optical sensor-based solutions in the upcoming years as well. Therefore, this research field also fosters the development of additional binding technologies, such as SHM and perimeter-security solutions.

Nowadays, most of the population understands the meaning of the term "optical internet." Optical fiber is used as a medium for the transmission of information to provide such a service to customers. Along with technology development, the number of applications of such optical fibers is increasing, and fiber-optic sensors are among these solutions.

The application of optical fibers in sensor designs was reported already in 1970. That became possible due to the invention of the laser in 1960 and the development of

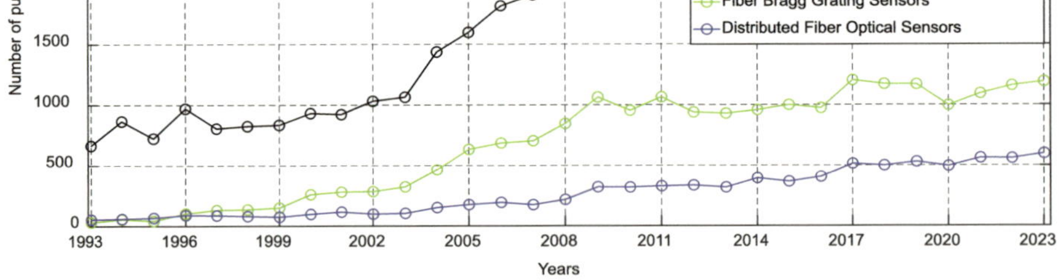

Figure 1.
Amount of research articles regarding FBG-based optical sensors indexed by SCOPUS database.

low-attenuation optical fibers in 1966 [6]. In the early 1990s, optical sensors, which could measure temperature, strain, pressure, and other parameters, were put into use. In turn, the beginning of the twenty-first century was characterized by the wide application of optical sensors, which measured temperature over the entire fiber length and allowed oil and gas companies to monitor their networks and timely repair damages. In the building construction sector, optical fiber sensors were used to monitor critical areas of the buildings and to plan timely repair works. In the meantime, process efficiency has been considered a growingly important issue, and optical fiber sensors have started playing an ever-increasing role.

Given the topicality of such FBG optical sensor utilization in various SHM applications in the next section of this work, an analysis of the most topical FOS FBG types and SHM applications is made, particularly discussing extrinsic and intrinsic optical fiber sensors. Fiber optical sensor classification is then made, as well as FBG sensor application category segregation is developed and evaluated. Further, descriptive analysis and design of FBG for the FOS solutions in SHM applications are shown. For this, potential FBG apodizations are studied and bandwidth, reflectivity, and side-lobe suppressions versus grating lengths are analyzed. Comparison analysis of such apodization profiles is then evaluated. Lastly, conclusions regarding the gained results are made.

2. Analysis of the most topical FOS FBG types and SHM applications

A lot of progress has been made since, and today, optical fiber sensors allow measuring changes in a wide range of parameters such as strain, pressure, temperature, movement, vibration, acceleration, rotation, moisture level, humidity, amperage, chemical composition, and many other parameters [6]. Development of those sensors was, to a great extent, promoted by the advantages optical fibers have over other kinds of sensors, for example, high sensitivity, immunity against electromagnetic interference, passive sensors, lightweight and compact size, chemical resistance, multiplexing opportunities, suitability for long-distance change monitoring, low-level induced attenuation [6], and wide range of measured chemical and physical parameters [7].

Optical fiber sensors can be classified into two large groups: intrinsic and extrinsic optical fiber sensors.

2.1 Extrinsic optical fiber sensors

In the scenario of extrinsic optical fiber sensors, optical signals are propagated outside the optical fiber, and a sensing element is needed at the end of the fiber. Materials that change their optical parameters, such as refraction index, absorption, fluorescence, reflection, etc., enable the production of a wide range of sensors depending on the monitored parameters. Nowadays, extrinsic optical fiber sensors still play an essential role in the general range of optical fiber sensors. Extrinsic optical fiber sensors use dual fiber connector sensors and Fabry–Pérot interferometers [8].

The dual fiber connector sensor is based on the principle in which the light beam of the transmitted optical signal interacts with a specific environment. The light beam is reflected from the said environment, and the receiver block analyzes how it affects the beam of light, factoring in observed parameters. Depending upon the observed parameter, the reflected signal can be further transmitted through the same or other fiber. The sensors of this variety can be utilized for monitoring temperature, displacement, pressure, and vibration [9].

Because of the relatively low manufacturing costs and compact size, the Fabry-Perrot interferometer is one of the first and most widely used extrinsic optical fiber sensors. The operational principle of the Fabry-Perrot interferometer is based upon two parallel reflective surfaces with a predetermined spacing between the two that is mainly filled with air, creating a cavity structure [10]. Drawbacks of this type of sensor are associated with coupling efficiency and calibration of the sensor [11]. Sensors of this kind are being deployed to detect pressure (e.g., gas, constructions from concrete and composites), single-point temperature points, and ultrasound monitoring [12, 13].

2.2 Intrinsic optical fiber sensors

Alternatively, intrinsic optical fiber sensors use optical fiber as a sensing element. Intrinsic optical fiber sensors can be classified into macrobending and microbending sensors, FBG sensors, as well as Rayleigh, Raman, and Brillouin scattering-based sensors. In the scenario of optical fiber scattering sensors (Rayleigh, Raman, and Brillouin), optical fiber is used both as a transmission medium and as a sensitive component. In sensors of such kind, optical fiber works as a sensor over the whole length.

Macrobending and micro-bending sensors are intensity-modulated sensors based on optical fiber bending. The micro-bending optical fiber sensor fundamentally consists of optical fiber inserted between two saw tooth structure deformation plates. Depending on the deformation plate, it can be modified to enable measurements of different parameters such as deformation, pressure, velocity, acceleration, weight, etc. The most substantial drawback of microlending sensors is high insertion loss (around 20 dB), which prevents them from being used in sensor network solutions where measurements are carried out at multiple points [9, 14].

For the macrobending sensor, the bending radius is many times larger than that of the microlending sensor, reaching 10–30 mm (depending on the utilized wavelength range). Macrobending sensors operate based on the principle of one or more bending loops. The attenuation increases with a reduction in the diameter of the optical fiber loop. Such sensors can be used as displacement (for monitoring buildings, railways, dams, embankments, and other objects), airflow, and temperature sensors. Given the fact that the attenuation of these sensors is many times lower than that of microbending sensors, it allows them to be multiplexed and deployed in an optical sensor network [15, 16].

In the use of optical fiber scattering (Rayleigh, Raman, Brillouin) based sensors, the optical fiber is used as both a transmission medium and a sensor. The whole length of the fiber acts as a sensor in this scenario. Scattering-based sensors do not require any modifications to their design to operate as sensors, but high-quality and high-sensitivity equipment is needed.

The power of the reflected signal due to the Rayleigh scattering is proportional to the optical power of the transmitted signal, which is approximately equal to λ^{-4} [17]. Energy is not transferred to the fiber during Rayleigh scattering. Furthermore, signal frequency is not altered as an effect of scattering, so it is considered elastic scattering, which is relative to the transmitted signal. The disadvantage of the mentioned sensors is introduced due to Rayleigh's scattered signal, which has low power (several dozen dB lower than the input signal) [17]. Regardless of the mentioned deficiency, such sensors are employed for monitoring temperature, deformations, vibrations [18], humidity [19], radiation [20], etc. It is important to note that the optical time domain reflectometer (OTDR) technology used to monitor optical lines (for monitoring insertion loss and impairments detection of optical lines) is based on Rayleigh scattering. If micro-bending and macrobending sensors are deployed in sensor network solutions, OTDR is also used to enable monitoring. Incorporating several fiber-optic sensor technologies and solutions into one (e.g., OTDR and FBG [21]) also seems interesting as this potentially provides an even more comprehensive range of monitoring services.

Optical fiber sensors based on stimulated Brillouin scattering (SBS) are used for temperature and deformation monitoring throughout the entire fiber length. Optical fiber Brillouin scattering sensors are based on inelastic scattering in which scattered waves experience frequency/wavelength shifts [22]. SBS-based sensors, depending on the realization principle and placement of continuous wave (CW) laser, can be further segregated into Brillouin optical time domain reflectometer (BOTDR) and Brillouin optical time domain analyzer (BOTDA) [17, 22, 23]. For BOTDA, the CW lasers and impulse signals propagate in different directions [17, 23]. This wavelength shift depends on temperature and deformation changes, allowing these devices to be used in sensor solutions. For BOTDR, the sensor's operational monitoring range is between 20 and 50 km. For BOTDA, it is possible to reach up to 200 km [17]. The spatial resolution for BOTDR is approximately 1 m (depending on the monitoring range), but for BOTDA, spatial resolution is 2 cm (for a monitoring range of 2 km) [24] and 2 m (for 150 km) [23].

Optical fiber sensors based on Raman's scattering are used to monitor temperature throughout the entire fiber length. The technological solution is broadly used for monitoring concrete structures, gas pipelines, and water supply systems [17, 25]. The operational monitoring range distance of the sensor can reach up to 37 km. Spatial resolution is closely dependent on monitoring distance, for example, at 1 km, spatial resolution is 1 cm, but at 37 km, spatial resolution is 17 m [23, 26, 27]. Studies indicate that optical fiber Raman scattering-based sensors can be used for temperature monitoring at distances 26 km away using SMF with a precision of 3 C and spatial resolution of 1 m [28]. Using graded-index few-mode fibers (GI-FMF), monitoring can be conducted for distances up to 25 km with a precision of 1°C and spatial resolution of 1,13 m [28–30].

The discovery of FBG significantly promoted the development of the optical fiber sensor and telecommunications industry. FBG is produced by modifying the refractive index within the core of an optical fiber (along the longitudinal axis). FBG has a wide range of applications, for example, in dispersion compensation, optical filters, optical fiber amplifiers, lasers, multiplexors, demultiplexers, sensors, and other solutions [31]. The reflected Bragg wavelength is sensitive to various physical parameters, so FBG can be used as an optical sensor to monitor and determine changes in physical parameters over time. FBG sensors as in quasi-distributed architecture, allow for simple multiplexing – using them in sensor network solutions (typically, approximately 100 sensors, but using CDM-WDM allows achieving the array of up to 1000 sensors) [32]. The spatial resolution of FBG sensors is equal to the grating length (typically, 2–10 mm, in rare configurations reaching up to 20 mm). The sampling frequency for standard optical sensors signal interrogation unit is up to 1 kHz; in rare cases, it reaches 5 kHz. FBG sensors can be used in various industries to measure and monitor temperature, strain, movement, pressure, vibration, and other physical parameters. FBG sensors tend to be most commonly used (2/3 of cases) in structural health monitoring (e.g., bridges, dams, roads, buildings, pipelines, etc.) [10, 23, 33].

2.3 Fiber optical sensor classification

To recognize the role and impact the optical sensors have on the monitoring and telecommunication sector, it is essential to classify them according to operational principle, spatial position, and field of application [6, 34, 35].

Based on operational principle or modulation type, sensors can be classified as [36–38]:

- Intensity-modulated sensors: Change of measured parameter is determined by intensity;
- Phase-modulated or interferometric sensors: Change of measured parameter is determined by the phase of signal;
- Polarization-modulated or polarimetric sensors: Change of measured parameter is determined by polarization change;
- Wavelength or spectrometric sensors: Change of measured parameter is determined by wavelength [6].

Based on the response to the measurement point sensors can be classified [36–39]:

- Point sensors: Measurement is performed at one point;
- Quasi-distributed scattered sensors: Measurements are performed at multiple points (sensor network);
- Distributed scattered sensors: The entire length of fiber operates as a sensor network.

Based on specific applications, field sensors can be classified as [36–38]:

- Physical (temperature, pressure, deformation, velocity, etc.);
- Chemical (pH level, gas analysis, spectroscopy, etc.);
- Biomedicine (blood flow, glucose level, etc.).

2.4 The application of FBG sensors

The application flowchart of FBG sensors, based on the parameters monitored by the sensors, is shown below in **Figure 2**. Most commonly, FBG sensors are used for strain and temperature monitoring of various roads [40–43], railways [44, 45], buildings [46–48], composite materials [49, 50], and healthcare facilities [51, 52].

For instance, when FBG FOS are applied for monitoring needs of the railways, motorways, buildings, and composite materials, commonly deformation sensors are mainly used for object's/infrastructure's structural monitoring [45–47, 50], whereas in

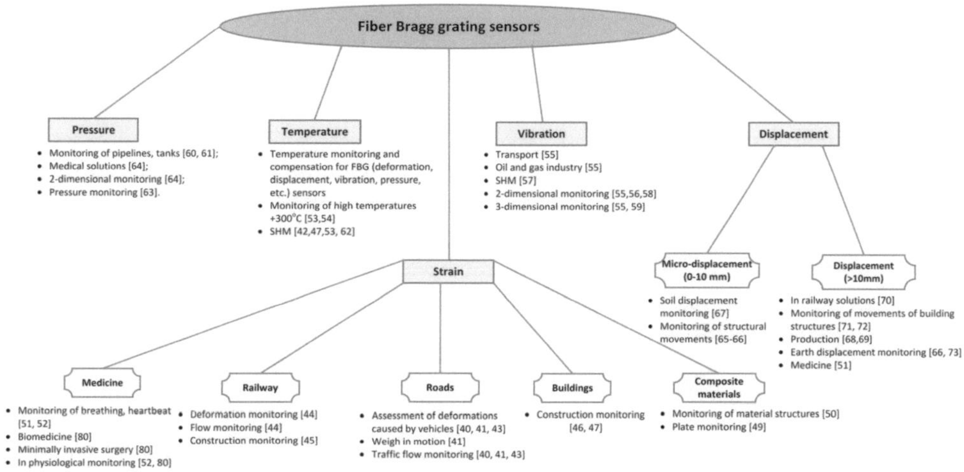

Figure 2.
Developed FBG FOS category model based on the main measured parameters and SHM, as well as security applications.

medicine [52, 53] for physiological monitoring of as well as monitoring of the respiratory and cardiac [51] related aspects.

Optical FBG temperature sensors can also compensate for induced temperature changes and their influence on the measurements of other physical parameters being monitored with such technology such as FBG strain, movement, vibration, pressure, etc. sensors. Temperature [54, 55] as a primary measurement can also be monitored with the FBGs, and such an approach can be utilized for structural health monitoring (SHM) and security [47, 54] solutions.

As for the FBG vibration sensors, such are used mainly in the oil and gas industry [56], SHM [57, 58], as well as in 2D and 3D vibration monitoring solutions [57, 59, 60].

In turn, FBG pressure sensors are used in pipeline [61], tank [62], and borehole [63] monitoring. Research results often report on additional types of developed pressure (2D and 3D) sensors [64, 65] for monitoring purposes.

FBG displacement sensors are classified by the range into micro-displacement (l = 0–10 mm) and movement (l > 10 mm) sensors. Micro-displacement sensors are less common as they monitor structure [66, 67] and soil [68] movement. In turn, movement sensors are used in industrial applications [69, 70], railway [71], and health care [61] solutions, as well as in the monitoring of building structures [72, 73] and ground movement [67, 74].

From our analysis, it is possible to evaluate that the FBG technology is a topical part of the FOS solutions within a variety of modern and industrial applications and monitoring solutions; therefore, it is important to address core component grating profile research and development to further improve and advance their applications within the SHM and perimeter monitoring solutions.

3. Analysis and design of FBG for the FOS solutions in SHM applications

By using optiwave system's "OptiGrating" software, the properties of the FBG were examined in the simulation environment. This software provides a variety of alternatives for the analysis and design of common grating-aided fibers and waveguides. Apodization type, index modulation, grating length, grating form, and period chirp are all adjusted during the construction of a Bragg fiber grating. The application can be used to do numerical simulations after such a device configuration is carried out. Solving coupled-mode equations, which explain how guided modes interact, is the foundation of these simulations. The transfer matrix method (TMM) is utilized to solve the coupled-mode equations. An overview of coupled-mode theory (CMT) for optical waveguides is presented by focusing on coupled optical waveguide analysis. It is applied to understand and examine the interplay of various light modes inside a waveguide. When there are many modes present in a waveguide, the waveguide's shape or refractive index variation may have an impact on how the modes interact. CMT uses mathematical formulas to depict the amplitudes of these various modes as they go through the waveguide. The energy transfer between modes and the amplitude changes with distance [75].

This concept makes the research of near resonance directed mode interaction possible. Furthermore, the coupled-mode equation may be solved using the TMM if the application has numerous gratings plus phase shifts. The study of nearly periodic gratings may also be done with TMM. The whole structure of the transfer matrix may be obtained by multiplying all the separate transfer matrixes, as shown by our researchers [76].

While FBG sensors have been widely employed for many applications, their implementation to enhance the performance of optical transmission links has transformed the communications sector. Researchers are using cascaded FBG (CFBG) structures to improve optical communication systems' performance by lowering the laser source's spectrum breadth and, therefore, the dispersion [77]. Because apodized grating offers excellent side-lobe suppression and improved filtering efficiency, apodized FBGs have been created and used extensively in optical fiber technology. An apodized periodic grating's reflection spectrum closely resembles the applied apodization profile's Fourier transform. Therefore, improved side-lobe suppression and higher grating performance are the outcomes of smooth and tight apodization [78].

Due to FBG's technological capabilities allowing for a certain reference wavelength inscription, light propagation in fiber is reflected in accordance with the Bragg wavelength. The combination of its component parts forms the entire reflected light, and this is only feasible if the Bragg condition given by the following equation is met: [79]

$$\lambda_B = 2 \cdot \Lambda \cdot n_{eff} \tag{1}$$

where n_{eff} is the effective refractive index of the fiber core, λ_B is the Bragg wavelength, and Λ is the grating period that establishes the separation between two consecutive grating planes. Along the grating length, the refractive index changes regularly as follows:

$$n_{eff}(z) = n_0 + f(z) \cdot \Delta n_{ac}(z) \cdot \cos\left(\frac{2\pi \cdot z}{\Lambda} + \theta(z)\right) + n_2 \cdot E^2(z) \tag{2}$$

where n_0 is the refractive index before grating inscription, Δn_{ac} is the refractive index modulation amplitude, z is the position, $f(z)$ is the apodization function, $\theta(z) = (2\pi/\Lambda)Cz^2$ is the chirp function, C is the chirp parameter, $E(z)$ is the electric field, and n_2 is the nonlinear Kerr coefficient [78].

In this work, the main apodization profiles we have investigated are given below:

- Gaussian [78]:

$$f(z) = \exp\left(-4 \cdot \ln (2) \cdot \left(\frac{2 \cdot z - L}{L}\right)^2\right) \tag{3}$$

- Tan hyperbola [80]:

$$f(z) = 0.5\left[1 + tanh\left(\beta\left(1 - 2\left(\frac{2z}{L}\right)^{\alpha}\right)\right)\right] \tag{4}$$

- Uniform [80]:

$$f(z) = 1 \tag{5}$$

- Nuttall:

$$f(z) = a_0 - a_1 \cdot \cos\left(2 \cdot \pi \cdot \frac{z}{L}\right) + a_2 \cos\left(4 \cdot \pi \cdot \frac{z}{L}\right) - a_3 \cdot \cos\left(6 \cdot \pi \cdot \frac{z}{L}\right) \tag{6}$$

where a_0 = 0.3635819, a_1 = 0.4891775, a_2 = 0.1365995, and a_3 = 0.0106411 are Nuttall coefficients, L is the grating length, and z is the coordinate of light propagation along the length of FBG [78].

In this practical experiment, where all OptiGrating simulations were carried out, the typical parameters of single-mode optical fiber with a core diameter of 8.2 μm and a cladding diameter of 125 μm were set. The central wavelength of each grating for FBG is set to 1550 nm.

In this experiment, different grating parameters were obtained from various FBG-reflected signal spectrums. In this experiment, we investigated FBG apodization performance to obtain an optimized reflection spectrum with high reflectivity (greater than 90%) and effective side-lobe suppression based on reflectivity level, full width at half maximum (FWHM) (less than 0.2 nm), and side-lobe suppression (SLS) (at least 20 dB) across various values of grating length, modulated refractive index (MRI), segment numbers, and different apodization profiles. In general, reflectivity (%), bandwidth (nm), and SLS (dB) are the three primary factors that need to be regulated while developing the FBGs. The research findings suggest highly efficient FBG grating apodizations that can be physically implemented for optical sensor networks and long-range (more than 40 km) structural health monitoring applications. Because reflectivity depends on both grating strength and length, the grating length may be adjusted to increase signal reflectivity. The following formula can be used to determine the signal's reflectivity [76]:

$$\text{Reflectivity}[\%] = 10^{\frac{P_{max}[dB]}{10}} \tag{7}$$

where P_{max} is the maximal signal power of the reflected signal spectrum.

SLS depends on the value of the side-lobe layer and the peak power value.

In this work, we have analyzed the signal spectrum at different grating lengths from 1 to 20 mm. While varying the grating length, different grating modulation indices were also set to get the optimal values of reflectivity, bandwidth, and side-lobe suppression of the spectral data. The modulation indices, which were set to determine the optimal values of the FBG sensors' key parameters are $\Delta n = 1.5 \times 10^{-5}$, $\Delta n = 2.5 \times 10^{-5}$, $\Delta n = 5 \times 10^{-5}$, $\Delta n = 15 \times 10^{-5}$ and $\Delta n = 25 \times 10^{-5}$. The refractive index modulation is consistently raised and uniformly decreased over the length of the FBG, which is an essential point to be emphasized based on the apodization approach. However, an increase in the overall grating length could be necessary to obtain a certain peak (reflectivity) of the reflected signal in this design.

3.1 Bandwidth versus grating length of FBG with apodizations

While testing the Gaussian apodization, FBG bandwidth decreases by increasing the grating length. For modulation indices 1.5×10^{-5}, 2.5×10^{-5}, and 5×10^{-5}, the bandwidth is the same, and the initial bandwidth - 1.072 nm, which is higher than the required threshold bandwidth, see **Figure 3**. However, increasing the grating length helped to reduce the bandwidth. It was less than 0.2 nm when the grating length was 7 mm, and it gradually decreased to 0.048 nm at 20 mm. For 15×10^{-5}, the bandwidth with the grating length of 1 mm was 1.072 nm, but then, with the increment of grating length, it started to decrease. It reached 0.176 nm at 7 mm grating length and 0.128 nm at 20 mm. For 25×10^{-5}, the bandwidth at 1 mm grating length was 1.088 nm, which was too high. After increasing the grating length, it fell less than 0.2 nm at 15 mm grating length (0.128 nm) and remained constant till 20 mm grating length.

For hyperbolic tangent apodization (**Figure 4**), the bandwidth for modulation indices 1.5×10^{-5} and 2.5×10^{-5} were almost the same. The bandwidth with the grating length of 1 mm was 0.72 nm, reaching 0.144 nm at a 5 mm grating length. From 5–20 mm grating length, the bandwidth was between 0.144 and 0.032 nm, which is below 0.2 nm. For 5×10^{-5}, the bandwidth was 0.144 nm at 5 mm grating length, dropping to 0.048 nm at 20 mm grating length. For 15×10^{-5}, the bandwidth was comparatively higher than 1.5, 2.5, and 5 modulation indices. Its lowest bandwidth at the grating length of 20 mm was 0.128 nm, which also fell within the required

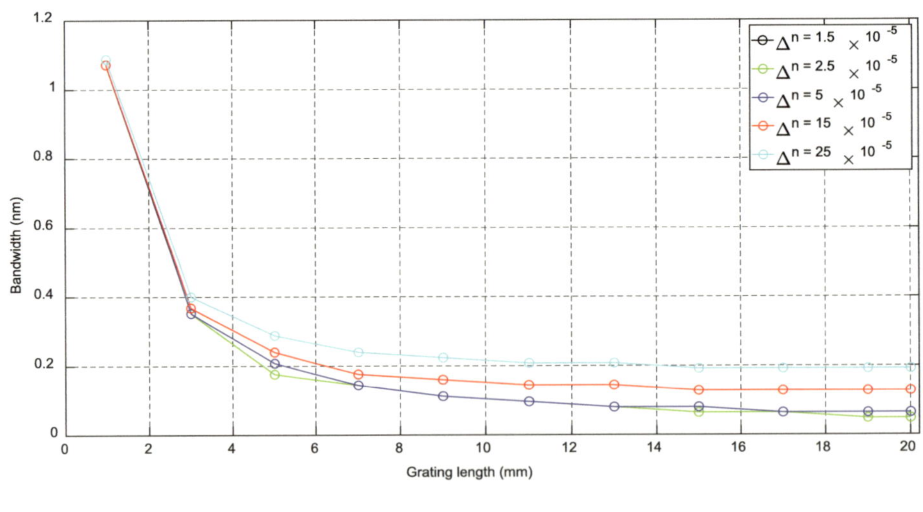

Figure 3.
Bandwidth versus grating length for Gaussian apodization profile simulation in Bragg grating profile.

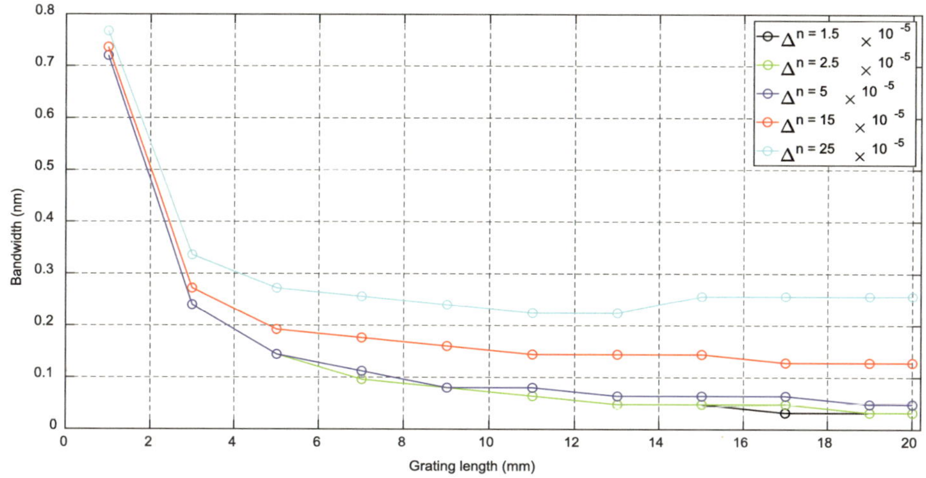

Figure 4.
Bandwidth versus grating length for hyperbolic tangent apodization profile simulation in Bragg grating profile.

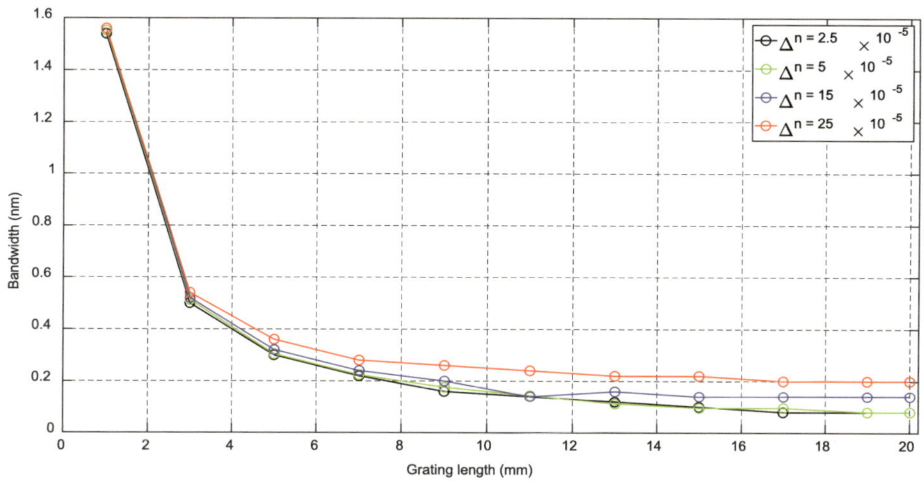

Figure 5.
Bandwidth versus grating length for Nuttall apodization profile simulation in Bragg grating profile.

range. For 25×10^{-5}, the bandwidth was high till the grating length was 20 mm. At 15 mm, the bandwidth increased to (0.256 nm) and remained almost constant till 20 mm grating length.

For Nuttall apodization (**Figure 5**), the bandwidth was too high initially for all the modulation indices (around 1.55 nm). For 2.5×10^{-5}, the bandwidth dropped to half when the grating length was increased. It reached <0.2 nm at 9 mm grating length and continued reducing till 20 mm grating length (0.08 nm at 20 mm grating length). For 5×10^{-5}, the bandwidth reached the required threshold at 9 mm grating length, with a value of 0.176 nm. The bandwidth was very small at 20 mm grating length, around 0.08 nm. For 15×10^{-5}, the bandwidth we achieved was 0.14 nm at 9 mm grating length, and it remained constant till 20 mm grating length. There was a slight rise at 13 mm (0.16 nm), then again, it dropped to 0.14 nm. For 25×10^{-5}, the bandwidth never dropped down less than 0.2 nm. It reached 0.2 nm at 17 mm grating length and remained constant till 20 mm grating length.

3.2 Reflectivity versus grating length of FBG with apodizations

The coherence between reflectivity, grating length, and grating modulation index of FBG gratings is evident in **Figures 6–8**. From the Gaussian apodization graph, with the gradual increment of grating length, reflectivity in % increases.

For the modulation index $\Delta n = 1.5 \times 10^{-5}$, the reflectivity is too low (the highest value is 5.64%). For grating length 1–5 mm, it remains close to 0%, which means Gaussian apodization requires higher modulation indices to achieve the defined threshold reflectivity. For 2.5×10^{-5}, it shows slightly better results compared to

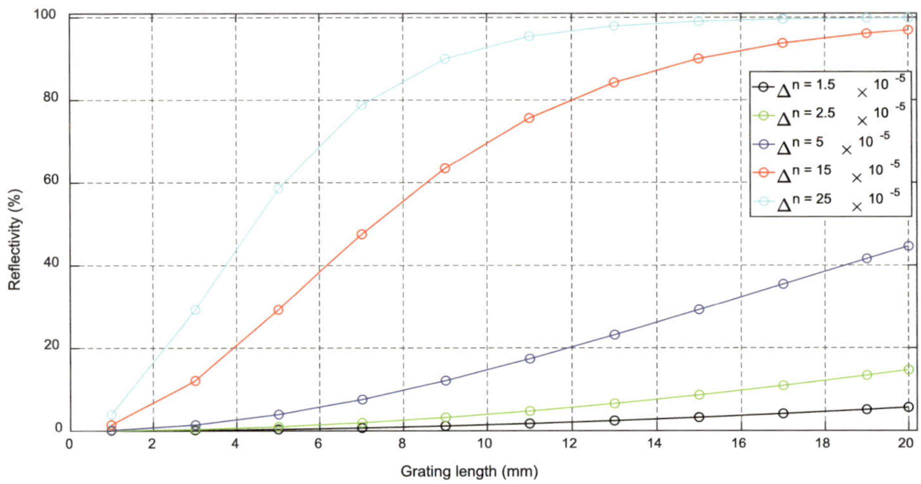

Figure 6.
Reflectivity versus grating length for Gaussian apodization profile simulation in Bragg grating profile.

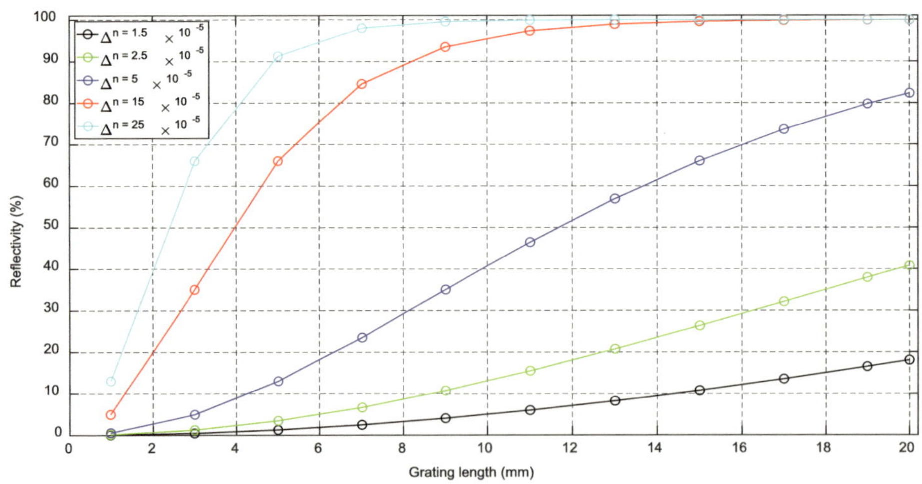

Figure 7.
Reflectivity versus grating length for hyperbolic tangent apodization profile simulation in Bragg grating profile.

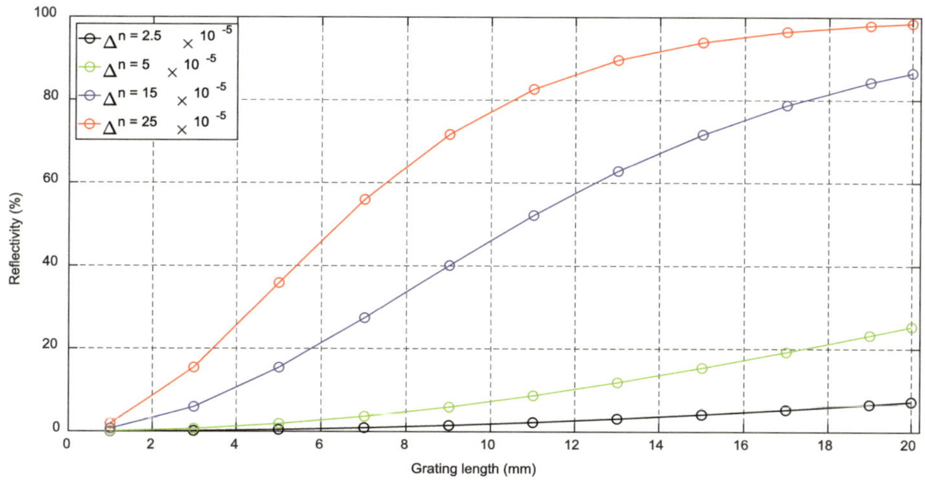

Figure 8.
Reflectivity versus grating length for Nuttall apodization profile simulation in Bragg grating profile.

1.5×10^{-5}. Still, it does not reflect the spectrum any closer to 90%. With a 20 mm grating length, its highest reflectivity is 14.66%. If we observe the reflectivity for 5×10^{-5}, it is noticeable that it shows better results compared to other modulation indices. The lowest reflectivity for a grating length of 1 mm is 0.16%; however, when the grating length is gradually increased, it rises to 44.6%, which is still not suitable for such FBG sensors. If we observe the reflectivity in % for modulation indices of 15×10^{-5} and 25×10^{-5}, they show improved results. For both modulation indices, the initial reflectivity was approximately 1%. However, the rise in grating length led them to achieve higher reflectivity (>90%). For 15×10^{-5}, the reflectivity reached 90% when the grating length was 17 mm, while for 25×10^{-5}, it reached 90% when the grating length was 11 mm and it continued to rise.

For hyperbolic tangent apodization (**Figure** 7), modulation indices of 1.5×10^{-5}, 2.5×10^{-5}, and 5×10^{-5} could not achieve the required reflectivity even after increasing the grating length (at 20 mm, the highest reflectivity was 18.07, 40.83 and 82.35%, respectively). Though the reflectivity increased gradually with the rise of grating length, it could not reach a reflectivity greater than 90%. For 15×10^{-5} and 25×10^{-5}, the recommended reflectivity was obtained with the increase in grating length. For 15×10^{-5}, it gained reflectivity of 93.49% at 9 mm grating length, and it increased to 99.95% at 20 mm grating length. For 25×10^{-5}, the required reflectivity was gained at a 5 mm grating length (91.29%), and it reached 100% at 15 mm.

For Nuttall apodization (**Figure 8**), a modulation index of 1.5×10^{-5} did not provide any significant result as it was too small to produce any values. Therefore, a modulation index of 2.5×10^{-5} was used as a simulation starting point. With the grating length of 1 mm, the reflectivity achieved was 0.02%. However, after increasing the grating length to 20 mm, it showed only 7.27%. For 5×10^{-5}, when the grating length was increased, the reflectivity also gradually increased, and at 20 mm, it reached 25.29%, which is also unacceptable. For 15×10^{-5}, the reflectivity with the grating length of 1 mm is 0.69%, but after setting up the grating length to 3 mm, it rose to 5%. It was increasing with the increment of grating length, and at 20 mm, it reached 86.51%. The reflectivity reached the aimed value after setting the modulation index to 25 × 10^{-5}. Although the initial value was 1.89%, the rise of grating length helped to achieve 93.88% (at 15 mm), it rose to 98.43% (at 20 mm).

3.3 Side-lobe suppression versus grating length of FBG apodizations

By increasing grating length, the side-lobe suppression of Gaussian apodization decreases. The achieved SLS was almost the same for modulation indices 1.5 × 10^{-5}, 2.5 × 10^{-5}, and 5 × 10^{-5}. The SLS was approximately 37 dB when the grating length was 1 mm, and after increasing the grating length, the SLS started dropping slightly. The highest drop (~3 dB) can be observed at 5 mm grating length (dropping from 37 to 34 dB). After that, it decreased relatively slowly and reached almost 32.5 dB (for 5 × 10^{-5} index) at the grating length of 20 mm.

For 15 × 10^{-5}, the SLS with 1 mm grating length was 36.98 dB, and after increasing the grating length, it continued dropping and reached 26.28 dB at 20 mm. For modulation index 25 × 10^{-5}, the initial SLS was 36.86 dB, and by increasing the grating length, the SLS fell to 21.93 dB at 20 mm, see **Figure 9**.

For hyperbolic tangent apodization (**Figure 10**), no modulation could reach the threshold value for SLS (at least 20 dB).

For 1.5 × 10^{-5} and 2.5 × 10^{-5}, the initial SLS value was almost 13.6 dB. After increasing the grating length, the SLS value did not decrease drastically. At 20 mm grating length, the SLS was 13.39 dB for a modulation index of 1.5 × 10^{-5} and 12.68 dB for 2.5 × 10^{-5}. For 5 × 10^{-5}, 15 × 10^{-5}, and 25 × 10^{-5}, the starting value with a 1 mm grating length was around 13 dB. However, after increasing the grating length, it changed dramatically. With 20 mm grating length, the SLS was 9.7 dB for Δn = 5 × 10^{-5}, 3.9 dB for Δn = 15 × 10^{-5}, and 2.42 dB for Δn = 25 × 10^{-5}, which is not suitable for FBG sensors and remote monitoring in such SHM solutions.

Figure 11 shows SLS vs. grating length for Nuttall apodization. The SLS is very high for all the modulation indices, around 59 to 60 dB. For 2.5 × 10^{-5} and 5 × 10^{-5}, the SLS fluctuated a lot within the range of 59 to 61 dB while increasing the grating

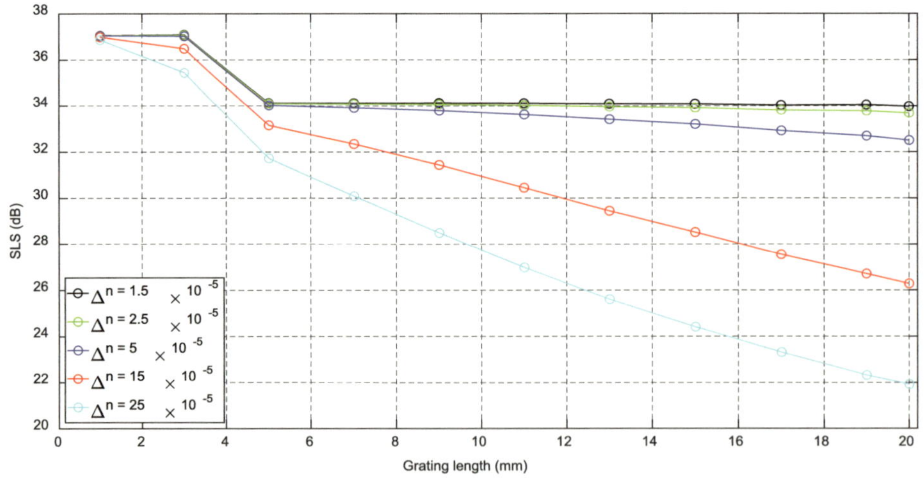

Figure 9.
Side-lobe suppression versus grating length for Gaussian apodization profile simulation in Bragg grating profile.

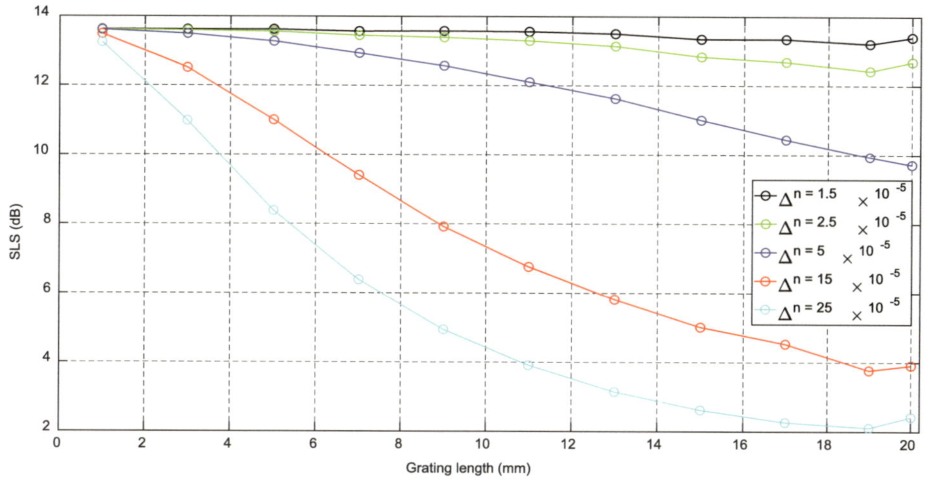

Figure 10.
Side-lobe suppression versus grating length for hyperbolic tangent apodization profile simulation in Bragg grating profile.

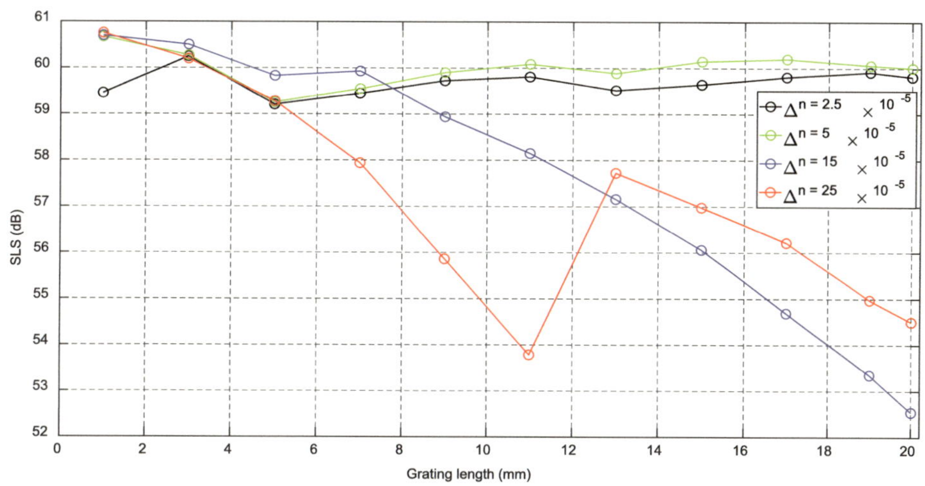

Figure 11.
Side-lobe suppression versus grating length for Nuttall apodization profile simulation in Bragg grating profile.

lengths. For 15×10^{-5}, the SLS was gradually decreasing, and it reached 52.54 dB at 20 mm grating length.

For 25×10^{-5}, SLS also decreased smoothly till 11 mm grating length, but at 13 mm grating length, it rose from 53 to 57 dB, and then again it continued to decrease and reached 54.45 dB at 20 mm. Such a pattern of sudden SLS increase (while increasing the grating length) and then again a decrease of the SLS while further increasing the grating length can be noticed due to the periodical function of SLS. Thus, when using even higher modulation indices than shown here, even more noticeable shifts in SLS values can be observed. Similar tendencies in a scientific journal article [76] and conference paper [81], where a modulation index of 2×10^{-4} [76] and 4×10^{-4} [81] is used and well seen when the SLS vs. grating length function is periodical.

3.4 Comparison analysis of apodization profiles

A modulation index of 15×10^{-5} was used for the comparison to more thoroughly examine and evaluate the FBG characteristics such as reflectivity (**Figure 12**), bandwidth (**Figure 13**), and SLS (**Figure 14**), and their relationship to the apodization profiles. In the comparison, the raised sinus apodization was additionally included, which showed the highest results in the previous study [76]. Choosing 15×10^{-5} modulation depends on how all the parameters produce the required result for different apodizations.

While analyzing the signal spectral reflectivity, it can be seen from **Figure 12** that by increasing the grating length, the reflectivity for all of the apodization profiles also

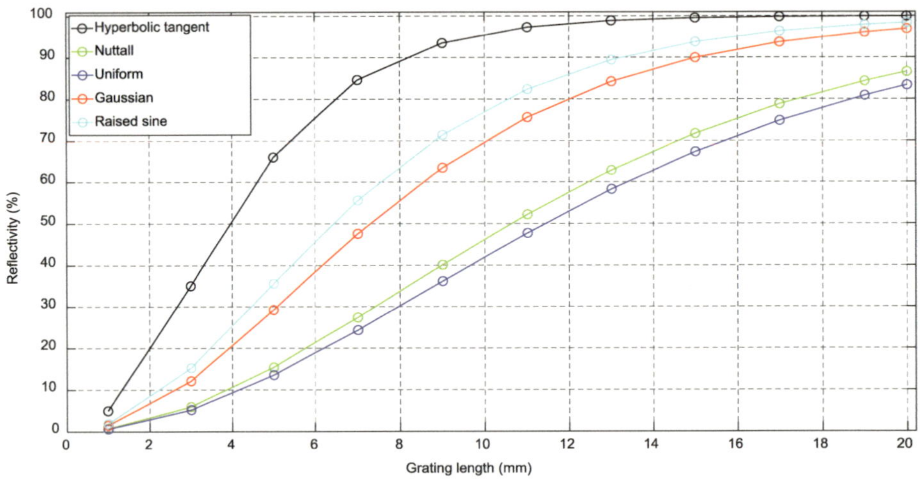

Figure 12.
Comparison of FBG apodization profiles regarding reflectivity and grating length.

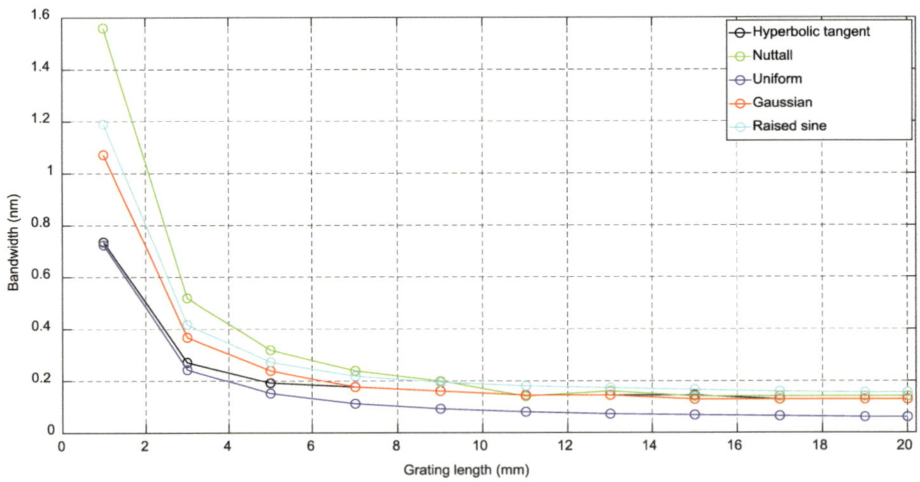

Figure 13.
Comparison for FBG apodization profiles regarding bandwidth and grating length.

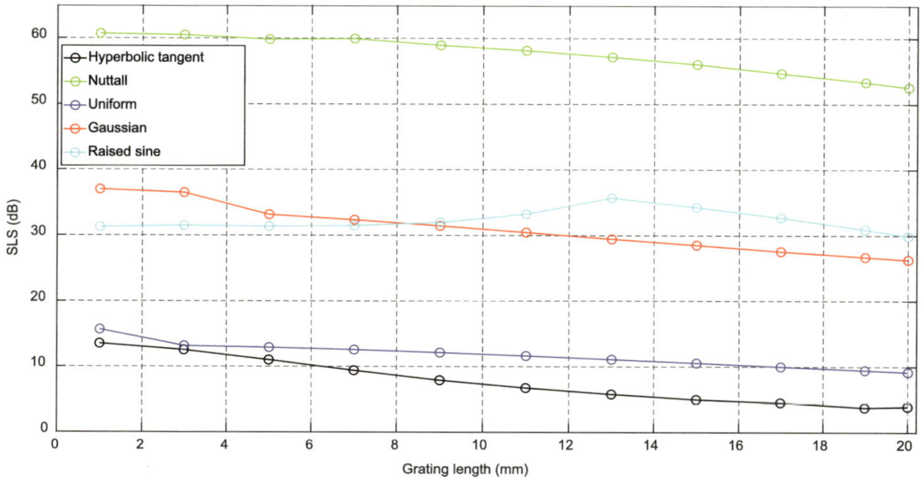

Figure 14.
Comparison for FBG apodization profiles regarding side-lobe suppression and grating length.

increases. The most rapid changes are noticed for hyperbolic tangent apodization within the 1–7 mm grating length interval. Hyperbolic tangent apodization shows the highest reflectivity results (99.95% at 20 mm) in the grating length range from 0 to 20 mm. Rapid changes are also observed for the sensors with grating lengths increasing from 1–11 mm in Gaussian and raised-sine apodizations. Meanwhile, the reflectivity increases almost linearly by increasing uniform (no apodization) and Nuttall apodization FBG grating length. So, choosing a higher number of grating lengths is important to get optimal reflectivity.

Figure 13 shows that by increasing the grating length, the reflected signal spectrum bandwidth decreases for all apodization profiles.

The most rapid changes are noticed for grating length 1–5 mm for hyperbolic tangent, 1–7 mm for Nuttall, raised sine, and Gaussian apodizations. Larger grating lengths (at least 7 mm) for FGBs should be selected for operating with a smaller bandwidth to preserve the optical frequency spectrum band. When grating lengths are selected between 1 and 7 mm, Gaussian provides the narrowest bandwidth, followed by hyperbolic tangent and Nuttall. As for the gratings between 9 and 20 mm in length, the narrowest bandwidth can be reached by applying the Gaussian and then hyperbolic tangent apodizations. The longer grating lengths for FGBs should be used to offer FBG with a smaller bandwidth (FWHM <0.2 nm) and high reflectivity (at least 90%).

Yet longer FBG gratings do not necessarily imply greater values for the SLS parameter (**Figure 14**). Therefore, it is essential to evaluate the most suitable FBG settings for each kind of grating. For hyperbolic tangent, the SLS is relatively low, around 13.48 dB. When the grating length increases, the SLS starts to decrease and at 20 mm, it reaches 3.91 dB. For Gaussian, the SLS falls abruptly for 1–5 mm grating length. After that, it decreases less, and at 20 mm grating length, the SLS is 26.28 dB. For Nuttall, the SLS is highest with a 1 mm grating length (60.69 dB). After that, it continuously decreases, and at 20 mm grating length, it reaches 52.54 dB.

4. Conclusions

In this study, we evaluated and analyzed the optical sensor and FBG, particularly integration solution possibilities in SHM applications. A clear and continuous increase in topicality trend could be observed. Additional analysis and segregation of the most topical FBG optical sensor types and SHM applications were conducted. This was done by segregating them based on the main measured parameters (such as pressure, temperature, vibration, displacement, and strain), as well as based on the modulation principle (intensity, phase, polarization, wavelength), and placement (point, quasi-distributed, distributed).

We also showed the simulation of multiple SHM potential grating apodization profiles (Gaussian, hyperbolic tangent, Nuttall). From the calculated data and set key performance indicators (reflectivity >90%, FWHM <0.2 nm, and SLS at least 20 dB), a comparison was made regarding various modulation indices, grating lengths, and apodization types. All of the three apodizations profiles with longer (9+ mm) grating lengths could reach a set 0.2 nm bandwidth. However, a slight advantage could be noticed when applying hyperbolic tangent and Gaussian apodizations. Reflectivity-wise, hyperbolic tangent apodization also shows a steeper rise while increasing the grating lengths, yet raised-sine and Gaussian apodizations can still reach the set 90% reflectivity rate. From the perspective of SLS, the increase of the grating length does not offer any such advantages, and all of the apodizations show a relatively stable trend. Yet, in this aspect, the hyperbolic tangent is the least effective. At the same time, Nuttall shows the highest dB ratio. Additionally, Gaussian and raised-sine also have an acceptable level of SLS.

While Nuttal apodization could be applied in configurations where efficient SLS management is of the most importance, bandwidth-wise and reflectivity-wise, it showed the least effective performance. In turn, the hyperbolic tangent was the least effective SLS-wise, yet it performed well regarding bandwidth and reflectivity. Gaussian and raised-sine apodization profiles, while possibly not being the most efficient in any of the measured parameters compared to other apodizations, still showed the possibility of reaching certain required thresholds. Therefore, these could also be evaluated for SHM and FBG optical sensor setups that require more stable and less risk-prone operations.

Acknowledgements

The chapter covering the literature review partly includes the content available in the doctoral thesis in the Latvian language by the same author: Janis Braunfelds. Characterization of development of the multifunctional fiber optical sensing system and its application in monitoring solutions. 2023. Riga Technical University. Available from: https://doi.org/10.7250/9789934229220.

Funding

This research was funded by the National Research Program within project No. VPP-EM-FOTONIKA2022/1–0001.

Data availability

The data used to support the findings of this study are available from the first author upon request.

Author details

Janis Braunfelds*, Ugis Senkans, Farjana Rahman, Nauris Silkans, Sandis Spolitis, Jurgis Porins and Vjaceslavs Bobrovs
Institute of Photonics, Electronics and Telecommunications, Riga Technical University, Riga, Latvia

*Address all correspondence to: janis.braunfelds@rtu.lv

References

[1] Jin Q, Liu Z, Bin J, Ren W. Predictive analytics of in-service bridge structural performance from SHM data mining perspective: A case study. Shock and Vibration. 2019;**2019**:6847053. DOI: 10.1155/2019/6847053

[2] Xu H, Li F, Zhao W, Wang S, Du Y, Bian C. A high precision fiber Bragg grating inclination sensor for slope monitoring. Journal of Sensors. 2019; **2019**:1354029, 7 pages. DOI: 10.1155/2019/1354029

[3] Budnicki D, Parola I, Szostkiewicz Ł, et al. All-fiber vector bending sensor based on a multicore fiberwith asymmetric air-hole structure. Journal of Lightwave Technology. 2020;**38**: 6685-6690. DOI: 10.1109/JLT.2020.3012769

[4] Liu Y, Zhou A, Yuan L. Multifunctional fiber-optic sensor, based on helix structure and fiber Bragg gratings, for shape sensing. Optics and Laser Technology. 2021;**143**:107327. DOI: 10.1016/j.optlastec.2021.107327

[5] Senkans U, Braunfelds J, et al. Research on FBG-based sensor networks and their coexistence with fiber optical transmission systems. Journal of Sensors. 2019;**2019**:6459387. DOI: 10.1155/2019/6459387

[6] Taylor & Francis Group. Optical Fiber Sensors – Advances Techniques and Applications. Boca Raton, United States: CRC Press; 2015. p. 545. DOI: 10.1201/b18074

[7] Advantages of optical sensor. Disadvantages of optical sensor. Available from: http://www.rfwireless-world.com/Terminology/Advantages-and-Disadvantages-of-Optical-Sensor.html [Accessed: December 5, 2022]

[8] Rajan G. Optical Fiber Sensors: Advanced Techniques and Applications. Boca Raton, United States: CRC Press; 2015, ISBN 9781482228250

[9] de Fátima M, Domingues F, Radan A. Optical Fiber Sensors for loT and Smart Devices. Vol. 101. Aveiro: Springer; 2017, ISBN: 978-3-319-47349-9

[10] Islam MR et al. Chronology of Fabry-Perot interferometer fiber-optic sensors and their applications: A review. Sensors. 2014;**14**(4):7451-7488. DOI: 10.3390/s140407451

[11] Lee CE et al. In-line fiber Fabry–Perot interferometer with high-reflectance internal mirrors. Journal of Lightwave Technology. 1992;**10**:105-108. DOI: 10.1109/ofs.1992.762972

[12] Chen P, Dai Y, et al. Cascaded-cavity Fabry-Perot interferometric gas pressure sensor based on Vernier effect. Sensors. 2018;**18**(11):3677. DOI: 10.3390/s18113677

[13] Li H, Li D, Xiong C, Si W, Fu C, Yuan P, et al. Low-cost, high-performance fiber optic Fabry–Perot sensor for ultrasonic wave detection. Sensors. 2019;**19**(2):406. DOI: 10.3390/s19020406

[14] Berthold JW. Historical review of microbend fiber-optic sensors. Journal of Lightwave Technology. 1995;**13**: 1193-1199. DOI: 10.1117/12.185034

[15] Zhikang Z, Yuangang L, et al. Fiber-optic macrobending-based temperature sensor with polyimide coating. Optica Sinica. 2022;**42**(7):0706008. DOI: 10.3788/AOS202242.0706008

[16] Roman M, Zhu C, et al. Distributed fiber-optic sensing with low bending loss

based on thin-core fiber. IEEE Sensors Journal. 2021;**21**(6):7672-7680. DOI: 10.1109/JSEN.2021.3050702

[17] Bao X, Chen L. Recent progress in distributed fiber optic sensors. Sensors. 2012;**12**:8601-8639. DOI: 10.3390/s120708601

[18] Peng F et al. Ultra-long high-sensitivity Φ-OTDR for high spatial resolution intrusion detection of pipelines. Optics Express. 2014;**22**(11):13804-13810. DOI: 10.1364/oe.22.013804

[19] Lenke P et al. Distributed humidity sensing based on Rayleigh scattering in polymer optical fibers. In: Fourth European Workshop on Optical Fibre Sensors. Porto, Portugal: International Society for Optics and Photonics; 2010. DOI: 10.1117/12.866279

[20] Rizzolo S et al. Radiation characterization of optical frequency domain reflectometry fiber-based distributed sensors. IEEE Transactions on Nuclear Science. 2016;**63**(3): 1688-1693. DOI: 10.1109/tns.2016.2527831

[21] Braunfelds J, Porins J, Spolitis S, Bobrovs V. Optical Sensor Interrogation Unit with Integrated Optical Time Domain Reflectometer Functionality. Riga, Latvia: Patent of the Republic of Latvia, LV15534; 2019

[22] Galindez-Jamioy CA et al. Brillouin distributed fiber sensors: An overview and applications. Journal of Sensors. 2012;**2012**:204121. DOI: 10.1155/2012/204121

[23] Barrias A et al. A review of distributed optical fiber sensors for civil engineering applications. Sensors. 2016; **16**:748. DOI: 10.3390/s16050748

[24] Dong Y, Zhang H, Chen L, Bao X. 2 cm spatial-resolution and 2 km range Brillouin optical fiber sensor using a transient differential pulse pair. Applied Optics. 2012;**51**:1229-1235. DOI: 10.1364/ao.51.001229

[25] Bolognini G et al. Raman-based fibre sensors: Trends and applications. Optical Fiber Technology. 2013;**19**(6):678-688. DOI: 10.016/j.yofte.2013.08.003

[26] Tanner MG, Dyer SD, Baek B, et al. High-resolution single-mode fiber-optic distributed Raman sensor for absolute temperature measurement using superconducting nanowire single-photon detectors. Applied Physics Letters. 2011;**99**:201110. DOI: 10.1063/1.3656702

[27] Park J, Bolognini G, Lee D, et al. Raman-based distributed temperature sensor with simplex coding and link optimization. IEEE Photonics Technology Letters. 2006;**18**:1879-1881. DOI: 10.1109/lpt.2006.881239

[28] Soto MA, Nannipieri T, et al. Raman-based distributed temperature sensor with 1 m spatial resolution over 26 km SMF using low-repetition-rate cyclic pulse coding. Optics Letters. 2011; **36**(13):2557-2559. DOI: 10.1364/ol.36.002557

[29] Liu Y, Ma L, Yang C, et al. Long-range Raman distributed temperature sensor with high spatial and temperature resolution using graded-index few-mode fiber. Optics Express. 2018;**26**(16): 20562-20571. DOI: 10.1364/OE.26.020562

[30] Lu P, Lalam N, et al. Distributed optical fiber sensing: Review and perspective. Applied Physics Reviews. 2019;**6**:041302. DOI: 10.1063/1.5113955

[31] Hill KO, Meltz G. Fiber Bragg grating technology fundamentals and overview.

Journal of Lightwave Technology. 1997; **15**(8):1263-1276. DOI: 10.1109/50.618320

[32] Götten M, Lochmann S, Ahrens A, et al. A CDM-WDM interrogation scheme for massive serial FBG sensor networks. IEEE Sensors Journal. 2022; **22**(12):11290-11296. DOI: 10.1109/JSEN.2021.3070446

[33] Wu T, Liu G, Fu S, Xing F. Recent progress of fiber-optic sensors for the structural health monitoring of civil infrastructure. Sensors. 2020;**20**:4517. DOI: 10.3390/s20164517

[34] Lecler S, Meyrueis P. Intrinsic optical fiber sensor. In: Fiber Optic Sensors. London, UK: IntechOpen; 2012. DOI: 10.5772/27079

[35] Kersey A, Dandridge A. Applications of fiber-optic sensors. IEEE Transactions on Components, Hybrids, and Manufacturing Technology. 1989;**13**: 472-478. DOI: 10.1109/ECC.1989.77791

[36] Braunfelds J, Spolitis S, et al. Demonstration of polarization optical-time-domain reflectometer for monitoring of optical fiber lines. In: 2022 Workshop on Microwave Theory and Techniques in Wireless Communications (MTTW). IEEE; 2022. pp. 9-12. DOI: 10.1109/MTTW56973.2022.9942583

[37] Li X, Yang C, Yang S, Li G. Fiber-optical sensors: Basics and applications in multiphase reactors. Sensors. 2012;**12**:12519-12544. DOI: 10.3390/s120912519

[38] Fidanboylu K, Efendioglu HS. Fiber optic sensors and their applications. In: Proceedings of 5th International Advanced Technologies Symposium (IATS'09), Karabuk, Turkey, 13–15 May 2009. 2009

[39] Güemes A et al. Structural health monitoring in composite structures by fiber-optic sensors. Sensors. 2018;**18**: 1094. DOI: 10.3390/s18041094

[40] Braunfelds J, Senkans U, Skels P, et al. FBG-based sensing for structural health monitoring of road infrastructure. Journal of Sensors. 2021;**2021**:8850368. DOI: 10.1155/2021/8850368

[41] Braunfelds J, Senkans U, Skels P, et al. Road pavement structural health monitoring by embedded fiber Bragg grating based optical sensors. Sensors. 2022;**22**:1-13. DOI: 10.3390/s22124581

[42] Braunfelds J, Senkans U, Skels P, et al. Fiber Bragg grating optical sensors for road infrastructure monitoring applications. In: Applied Industrial Optics. Dublin, Ireland: Optica Publishing Group; 2022. pp. 1-2. DOI: 10.1364/AIO.2022.W1A.2

[43] Yuksel K, Kinet D, Chah K, Caucheteur C. Implementation of a mobile platform based on fiber Bragg grating sensors for automotive traffic monitoring. Sensors. 2020;**20**:1567. DOI: 10.3390/s20061567

[44] Van Esbeen B et al. Smart railway traffic monitoring using fiber Bragg grating strain gauges. Sensors. 2022;**22**: 3429. DOI: 10.3390/s22093429

[45] Cocking S, Alexakis H, DeJong M. Distributed dynamic fibre-optic strain monitoring of the behaviour of a skewed masonry arch railway bridge. Journal of Civil Structural Health Monitoring. 2021; **11**(4):989-1012. DOI: 10.1007/s13349-021-00493-w

[46] Li C, Tang J, et al. FBG arrays for quasi-distributed sensing: A review. Photonic Sensors. 2021;**11**(1):91-108. DOI: 10.1007/s13320-021-0615-8

[47] Juraszek J, Antonik-Popiołek P. Fibre optic FBG sensors for monitoring of the temperature of the building envelope. Materials (Basel). 2021;**14**(5):1207. DOI: 10.3390/ma14051207

[48] Čápová K, Velebil L, et al. Environmental testing of a FBG sensor system for structural health monitoring of building and transport structures. Procedia Structural Integrity. 2019;**17**: 726-733. DOI: 10.1016/j.prostr.2019.08.097

[49] Kefal AI, Tabrizi E, Tansan M, et al. An experimental implementation of inverse finite element method for real-time shape and strain sensing of composite and sandwich structures. Composite Structures. 2021;**258**:113431. DOI: 10.1016/j.compstruct.2020.1131

[50] Falcetelli F, Martini A, et al. Strain modal testing with fiber Bragg gratings for automotive applications. Sensors. 2022;**22**:946. DOI: 10.3390/s22030946

[51] Lo PD et al. Wearable system based on flexible FBG for respiratory and cardiac monitoring. IEEE Sensors Journal. 2019;**19**(17):7391-7398. DOI: 10.1109/JSEN.2019.2916320

[52] Lo PD et al. Fiber Bragg gratings for medical applications and future challenges: A review. IEEE Access. 2020; **8**:156863-156888. DOI: 10.1109/ACCESS.2020.3019138

[53] Presti DL et al. Smart textile based on 12 fiber Bragg gratings array for vital signs monitoring. IEEE Sensors Journal. 2017;**17**(18):6037-6043. DOI: 10.1109/JSEN.2017.2731788

[54] Dutta C, Kumar J, Das TK, Sagar SP. Recent advancements in the development of sensors for the structural health monitoring (SHM) at high-temperature environment: A review. IEEE Sensors Journal. 2021;**21**(14): 15904-15916. DOI: 10.1109/JSEN.2021.3075535

[55] Ma S, Xu Y, Pang Y, et al. Optical fiber sensors for high-temperature monitoring: A review. Sensors. 2022;**22**: 5722. DOI: 10.3390/s22155722

[56] Li T, Guo J, Tan Y, Zhou Z. Recent advances and tendency in fiber Bragg grating-based vibration sensor: A review. IEEE Sensors Journal. 2020;**20**(20): 12074-12087. DOI: 10.1109/JSEN.2020.3000257

[57] Munendhar P, Khijwania SK. Two dimensional fiber Bragg grating based vibration sensor for structural health monitoring. AIP Conference Proceedings. 2013;**1536**:1324-1326. DOI: 10.1063/1.4810731

[58] Antunes P, Travanca R, Rodrigues H, et al. Dynamic structural health monitoring of slender structures using optical sensors. Sensors. 2012;**12**:6629-6644. DOI: 10.3390/s120506629

[59] Zhang Y et al. 2-D medium–high frequency fiber Bragg gratings accelerometer. IEEE Sensors Journal. 2017;**17**(3):614-618. DOI: 10.1109/JSEN.2016.2636453

[60] Xiong L, Jiang G, Guo Y, Liu H. A three-dimensional fiber Bragg grating force sensor for robot. IEEE Sensors Journal. 2018;**18**(9):3632-3639. DOI: 10.1109/JSEN.2018.2812820

[61] Wang J, Zhao L, Liu T, et al. Novel negative pressure wave-based pipeline leak detection system using fiber Bragg grating-based pressure sensors. Journal of Lightwave Technology. 2017;**35**(16): 3366-3373. DOI: 10.1109/JLT.2016.2615468

[62] Ko P-L, Chuang K-C, Ma C-C. A fiber Bragg grating-based thin-film sensor for measuring dynamic water pressure. IEEE Sensors Journal. 2018; **18**(18):7383-7391. DOI: 10.1109/JSEN.2018.2857561

[63] Wang Q, Zhang L, Sun C, Yu Q. Multiplexed fiber-optic pressure and temperature sensor system for down-hole measurement. IEEE Sensors Journal. 2008;**8**(11):1879-1883. DOI: 10.1109/JSEN.2008.2006253

[64] Li X et al. An FBG pressure sensor based on spring-diaphragm elastic structure for ultimate pressure detection. IEEE Sensors Journal. 2022;**22**(3): 2213-2220. DOI: 10.1109/JSEN.2021.3136212

[65] Kanellos GT, Papaioannou G, et al. Two dimensional polymer-embedded quasi-distributed FBG pressure sensor for biomedical applications. Optics Express. 2010;**18**:179-186. DOI: 10.1364/oe.18.000179

[66] Xiong L, Guo Y, Jiang G, Jiang L, Zhou X. Fiber Bragg grating displacement sensor with high measurement accuracy for crack monitoring. IEEE Sensors Journal. 2019; **19**:10506-10512. DOI: 10.1109/jsen.2019.2930761

[67] Bonopera M. Fiber-Bragg-grating-based displacement sensors: Review of recent advances. Materials (Basel). 2022;**15**(16):5561. DOI: 10.3390/ma15165561

[68] Hong C, Zhang Y, Yang Y, Yuan Y. A FBG based displacement transducer for small soil deformation measurement. Sensors and Actuators A: Physical. 2019; **286**:35-42. DOI: 10.1016/j.sna.2018.12.022

[69] Thomas J et al. Temperature compensated FBG displacement sensor for long-range applications. IEEE Sensors Letters. 2020;**4**:1-4. DOI: 10.1109/LSENS.2019.2959377

[70] Lu Z, Hong C, Zhang Y, Su D, Fu Y. Development of an FBG sensor for measuring large range and multi-directional settlement. IEEE Access. 2019;**7**:107669-107677. DOI: 10.1109/ACCESS.2019.2932774

[71] Li H, Xu G, Gui X, Liang L, Li Z. An FBG displacement sensor in deformation monitoring of subway floating slab. IEEE Sensors Journal. 2021; **21**(3):2963-2971. DOI: 10.1109/jsen.2020.3022466

[72] Li S, Zhongxin S, Chunwei Z, Yansheng S. A pitman style fiber Bragg grating displacement sensor based on wedge cavity structure. Mechanisms and Machine Science. 2019; **72**:313-324. DOI: 10.1007/978-3-030-17677-8_24

[73] Kim SW, Yun DW, Park DU, Chang SJ, Park JB. Vehicle load estimation using the reaction force of a vertical displacement sensor based on fiber Bragg grating. Sensors. 2022;**22**: 1572. DOI: 10.3390/s22041572

[74] Alias MA, Ismail MF, et al. A high-precision extensometer system for ground displacement measurement using fiber Bragg grating. IEEE Sensors Journal. 2022;**22**:8509-8521. DOI: 10.1109/jsen.2022.3159850

[75] Optiwave. OptiGrating User's Reference and Tutorials. Canada: Optiwave; 2008. 208 p

[76] Braunfelds J et al. Designing of fiber Bragg gratings for long-distance optical fiber sensing networks. Modelling & Simulation in Engineering. 2022;**2022**: 1-14. DOI: 10.1155/2022/8331485

[77] Bhaskar CVN, Pal S, Pattnaik PK. Performance enhancement of optical communication system with cascaded FBGs of varying lengths. Journal of Optics. 2023;**25**:1-13. DOI: 10.1088/2040-8986/ad0def

[78] Chaoui F, Aghzout O, Chakkour M. Apodization optimization of FBG strain sensor for quasi-distributed sensing measurement applications. Active and Passive Electronic Components. 2016; **2016**:1-8. DOI: 10.1155/2016/6523046

[79] Shekar PVR, Latha DM, Kumari K. Optimal parameters for fber Bragg gratings for sensing applications: A spectral study. SN Applied Sciences. 2021;**3**(45):1-12. DOI: 10.1007/s42452-021-04650-0

[80] Palumbo G et al. Analysis and design of chirped fiber Bragg grating for temperature sensing for possible biomedical applications. IEEE Photonics Journal. 2018;**10**:1-17. DOI: 10.1109/JPHOT.2018.2829623

[81] Dwivedi KM, Trivedi G, Khijwania SK. Theoretical study and optimization of apodized fiber Bragg grating for single and quasi-distributed structural health monitoring applications. In: 30th International Conference Radioelektronika. Bratislava, Slovakia: Radioelektronika; 2020. DOI: 10.1109/RADIOELEKTRONIKA49387.2020.9092399

Chapter 2

Theoretical Investigation of Distributed Fiber Optic Sensing: Brillouin Stimulated Scattering

Alireda Aljaroudi and Ahmed Aljaroudi

Abstract

The primary function of monitoring systems is to ensure the operational integrity and safety of the system being monitored. Continuous monitoring provides timely information on the state of the monitored system as well as early warning of anomalies that might lead to catastrophic failures. One of the leading sensing technologies that is gaining wide acceptance in the industry is fiber optic-based sensing technology. This technology enables the fiber optic cable to act as a sensor, providing continuous sensing with wide sensing coverage and advanced warning capabilities in real-time. When utilized in a distributed configuration, it detects changes in vibration, strain, and temperature along the monitored object. The strain on an offshore crude pipeline may indicate the presence of fractures; recognizing them early can prevent structural collapse, which might lead to a leak and, eventually, an oil spill, inflicting environmental damage. This chapter presents a theoretical analysis of fiber optic-based sensing and its practical applications.

Keywords: stimulated Brillouin scattering, stimulated Brillouin gain, stimulated Brillouin loss, distributed fiber optic sensing, Brillouin optical time domain analysis, temperature and strain sensing

1. Introduction

One of the most promising technologies that can provide ongoing monitoring is distributed fiber optic sensing. It can perform continuous sensing along the entire length of the monitored structure with great accuracy and precision. It enables the fiber optic cable to function as a sensor, providing greater sensing coverage and continuous real-time sensing capabilities. Moreover, it provides increased temperature accuracy and exceptional spatial resolution. When the optical laser propagates through the fiber, it gets scattered back into three different spectral forms with different frequencies and intensities. The spectral forms are called Rayleigh, Raman, and Brillouin scattering [1]. The scattering is created due to impurities or changes in composition and interaction of the laser light with the molecules of the fiber.

The scattering causes a frequency shift of the signals returning to the source and is dependent on the strain and temperature of the sensing fiber [2]. Thus, it can be used to reveal the change of temperature and strain that may take place along the fiber, or any object attached to the fiber cable.

Due to intrinsic impurities and microscopic flaws in manufacturing, scattering mechanisms are always present in the fiber. The scattering of light in an optical fiber is caused by variations in the fiber's optical characteristics, such as density or refractive index. As mentioned above, scattering occurs in three different forms, Rayleigh, Raman, and Brillouin. Rayleigh scattering does not cause a frequency shift of the scattered light as it is an elastic process and is insensitive to changes in the sensing fiber material. Conversely, Raman and Brillouin scattering are inelastic processes and are sensitive to material properties.

Raman scattering is sensitive to temperature only and occurs in the forward and backward directions. While Brillouin scattering occurs in the backward scattering direction only and is sensitive to both temperature and strain. These two processes involve three waves that are of importance in the sensing regime. The three waves are the incident and the scattered waves as well as the thermal molecular vibration within the fiber.

There are two scattering schemes associated with this type of scattering, these are the spontaneous Brillouin and the stimulated Brillouin scattering [3]. When the incident light does not impose any changes on the optical material properties, we have a spontaneous scattering. On the other hand, when the incident light imposes a change, we have a stimulated scattering.

Along the way, a vibration of the fiber molecules takes place traveling at the acoustic speed which causes variation in density. Accordingly, the refractive index may change causing the light to scatter back to the source. This vibration is referred to as phonons. Basically, it is the scattering of light from sound waves in the fiber.

For distributed sensing, it is more attractive to use a Brillouin-based sensing technique rather than a Raman-based technique. This is because Brillouin scattering can perform both temperature and strain sensing. Moreover, Brillouin-based sensing outperforms Raman sensing in terms of sensing range, resolution and measuring time [1]. Therefore, the focus of this chapter is on the Brillouin scattering technique.

2. The stimulated Brillouin scattering

The Stimulated Brillouin Scattering (SBS) results from the interaction between two optical waves propagating in opposite directions, refer to **Figure 1**. As a result of this interaction, an acoustic wave will emerge that causes changes in the refractive index of the fiber resulting in coupling of the two counter-propagating optical waves.

Assuming steady state conditions, we will have two coupled wave equations that govern the scattering process. Assuming the fiber loss, α is the same for both waves, then the coupled wave equations can be expressed as [4]:

$$\frac{dI_P}{dz} = -g_B I_P I_{CW} - \alpha I_P \tag{1}$$

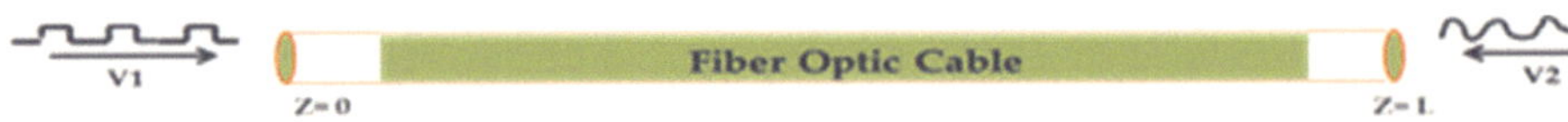

Figure 1.
Illustration of a simplified Brillouin system.

$$\frac{dI_{CW}}{dz} = -g_B I_P I_{CW} + \alpha I_{CW} \tag{2}$$

where I_P and I_{CW} are the input pump intensity and probe intensity, respectively, α is the fiber optic cable attenuation coefficient, g_B is the Brillouin gain and z is the distance.

2.1 Stimulated Brillouin gain

SBS is categorized into two forms, the Brillouin gain spectrum and the Brillouin loss spectrum. Brillouin gain operates by launching a pulsed laser ν_1, the pump at one end of the fiber and a continuous wave (CW) probe laser ν_2 into the other end of the fiber where pump frequency is greater than probe frequency. When the frequency of the pulsed laser (the pump laser) is higher than the frequency of the probe laser by the Brillouin frequency shift, ν_B, then there will be a transfer of energy from the pump to the probe. In this case, the probe will experience gain [5]. Eq. (3) presents this relationship between the two waves as:

$$\nu_1 = \nu_2 + \nu_B \tag{3}$$

For the Brillouin loss, the frequencies are reversed (probe frequency > pump frequency). In this case, there is a transfer of energy from the probe to the pump and the probe experiences loss. Eq. (4) expresses this relationship as:

$$\nu_2 = \nu_1 + \nu_B \tag{4}$$

Based on the above discussion, the probe signal is amplified in the Brillouin gain process, and it is reduced in the Brillouin loss process. In the SBS configuration, we want to measure the signal coming toward the source. In this case, we want to know the intensity or the power level of the probe signal.

2.2 Stimulated Brillouin loss

The focus of this chapter is on Brillouin loss. Accordingly, the sign of the gain in Eqs. (1) and (2) should be reversed [6].

$$\frac{dI_P}{dz} = g_B I_P I_{CW} - \alpha I_P \tag{5}$$

$$\frac{dI_{CW}}{dz} = g_B I_P I_{CW} + \alpha I_{CW} \tag{6}$$

Integrating Eq. (5), we get:

$$\int_0^z \frac{dI_P}{I_P} = \int_0^z (g_B I_{CW} - \alpha)dz \tag{7}$$

At the time the pump laser is launched at point 0 ($z = 0$), there will be no gain or loss, then Eq. (7) becomes:

$$ln\, I_P(z) - ln\, I_P(0) = -\alpha z \tag{8}$$

$$\left(\frac{I_P(z)}{I_P(0)}\right) = exp\,(-\alpha z) \tag{9}$$

$$I_P(z) = I_P(0)\,exp\,(-\alpha z) \tag{10}$$

Eq. (10) calculates the intensity before the interaction between the two lasers.

Integrating Eq. (6) over the entire length (from z = L to z = 0), we get the intensity of the probe laser at distance z = 0:

$$\frac{dI_{CW}}{dz} = g_B I_P I_{CW} + \alpha I_{CW} = \quad \int_L^0 \frac{dI_{CW}}{I_{CW}} = \int_L^0 (g_B I_P(z) + \alpha)dz \tag{11}$$

$$\frac{dI_{CW}}{dz} = g_B I_P I_{CW} + \alpha I_{CW} = \quad \int_L^0 \frac{dI_{CW}}{I_{CW}} = \int_L^0 (g_B I_P(0)\,exp\,(-\alpha z) + \alpha)dz \tag{12}$$

$$ln\left(\frac{I_{CW}(0)}{I_{CW}(L)}\right) = \left(\frac{g_B I_P(0)}{\alpha}\right)(exp\,(-\alpha L) - 1) - \alpha L \tag{13}$$

$$I_{CW}(0) = I_{CW}(L)\,exp\left[\left(\frac{g_B I_P(0)}{\alpha}\right)(exp\,(-\alpha L) - 1) - \alpha L\right] \tag{14}$$

Eq. (14) calculates the intensity of the probe laser at point 0 (z = 0) [6].

As the signal propagates along the fiber, the power of the signal drops due to the fiber optic cable attenuation. However, the power of the signal becomes constant at a specific length of the fiber, which is known as the effective length. This effective length can be expressed as [4]:

$$L_{eff} = \frac{1}{\alpha}(1 - \exp(-\alpha L)) \tag{15}$$

As an example, let us assume the attenuation, α of a fiber optic cable is $0.2\ dB/_{km}$ and the length of the fiber optic cable is 20 *km*, then effective length is calculated to be 13 *km*. It must be noted that the first step in this example is to convert $\alpha\ dB/_{km}$ to α/km using Eq. (16) [4]. The derivation for this equation is presented in Appendix A.

$$\alpha_{dB/_{km}} = 4.343_{\alpha/_{km}} \tag{16}$$

Using Eq. (15) in Eq. (14) yields:

$$I_{CW}(0) = I_{CW}(L)\,exp\,[-g_B I_P(0) L_{eff} - \alpha L] \tag{17}$$

The power of the signal is not equally distributed over the cross-sectional area of the fiber optic cable; hence, it is appropriate to use the effective area, A_{eff} instead of the cross-sectional area. The effective area is the region where intensity of light does not exhibit nonlinear behaviors. It can be expressed in terms of the intensity, I, and power, P, as [4, 6]:

$$A_{eff} = \frac{P}{I}, \quad I = \frac{P}{A_{eff}} \tag{18}$$

Substituting Eq. (18) into Eq. (17) yields:

$$P_{CW}(0) = P_{CW}(L)\,exp\left[\left(-\frac{g_B P_P(0) L_{eff}}{A_{eff}}\right) - \alpha L\right] \tag{19}$$

$P_{CW}(0)$ is the power spectrum of the probe signal at $z = 0$. From Eq. (10), pump power at z can be expressed as:

$$P_P(z) = P_P(0)\exp(-\alpha z) \tag{20}$$

2.3 Brillouin spectrum

The acoustic waves decay exponentially, and the Brillouin gain spectrum takes the shape of Lorentzian curve [6, 7]. It is defined as:

$$g_{B(\nu)} = g_{B0}\frac{\left(\frac{\Delta\nu_B}{2}\right)^2}{\left(\frac{\Delta\nu_B}{2}\right)^2 + (\nu - \nu_B)^2} = \frac{g_{B0}}{1 + \frac{4(\nu-\nu_B)^2}{\Delta\nu_B{}^2}} \tag{21}$$

where g_B is the Brillion gain, ν is the probe-pump frequency difference at location z along the fiber, g_{Bo} is the Brillouin gain that occurs when the frequency difference ν is equal to Brillouin frequency shift $\nu_{B,}$ and $\Delta\nu_B$ is the Brillouin bandwidth. A typical Brillouin gain spectrum is shown in **Figure 2** which indicates that the peak power or Brillouin gain occurs at the Brillouin frequency shift, ν_B. As the figure illustrates, the shape of the spectrum may approximate Lorentzian distribution.

At the maximum gain, we can get the Brillouin frequency shift ν_B, from which we can determine the temperature or strain at a particular position along the fiber optic cable or any object that is attached or bonded to the fiber optic cable.

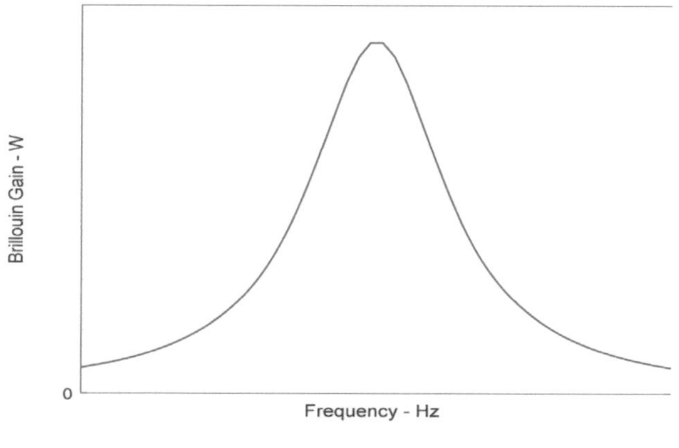

Figure 2.
Illustration of a typical Brillouin gain spectrum.

3. Distributed Brillouin sensing techniques

Brillouin sensing techniques include Brillouin Optical Time Domain Reflectometry (BOTDR), Brillouin Optical Time Domain Analysis (BOTDA), Brillouin Optical Frequency Domain Analysis (BOFDA), Brillouin Optical Correlation-Domain Analysis (BOCDA), and Brillouin Echo Distributed Sensing (BEDS). The BOTDA technology is extensively utilized as a monitoring tool in the industry. It provides excellent temperature and strain-detecting capability [8].

3.1 Brillouin optical time domain analysis (BOTDA)

As previously explained, BODTA, when utilized in the stimulated Brillouin loss scheme, operates by firing lasers in two opposing directions, one pulsed and one continuous, as seen in **Figure 1**. The difference in frequency between the two lasers may be utilized to monitor strain and temperature along the fiber [1]. This frequency difference is frequently referred to as the Brillouin frequency shift because it happens in the reverse direction. It is the consequence of dispersed light, which is affected by material temperature and tension. This is one of the primary benefits of employing fiber optics as a distributed sensor to detect and quantify strain and temperature variations along the monitored structure.

The frequency shift is given by, [4]:

$$\nu_B = \frac{2n\mathrm{V}_a \, sin\left(\frac{\theta}{2}\right)}{\lambda_i} \tag{22}$$

Since the scattering happens only in the backward direction, $\theta = 180^0$, then the frequency shift will be:

$$\nu_B = \frac{2n\mathrm{V}_a}{\lambda_i} \tag{23}$$

ν_B is the Brillouin frequency shift, V_a is the acoustic velocity of the phonons (approximately 5800 m/s), n is the refractive index of the fiber, λ_i is the wavelength of the incident light. A photon is a particle of light while a phonon is a particle of sound, it is a unit of vibration energy. Both photons and phonons are the smallest amount of energy that a wave can carry. Assuming a wavelength of 1300 nm and a refractive index of 1.5, Eq. (23) yields a Brillouin frequency shift of 13,384.6 MHz.

3.2 Fiber optic distributed sensing based on Brillouin scattering

A linear relationship exists between the frequency shift and strain. Similarly, a linear relationship exists between the frequency shift and the temperature [5, 6, 9].

$$\nu_B - \nu_0 = \alpha_\varepsilon \Delta\varepsilon + \alpha_T \Delta T \tag{24}$$

ν_B is the Brillouin frequency shift, ν_0 is the reference Brillouin frequency shift at no strain and ambient temperature in MHz, α_ε is the strain coefficient expressed in MHz/ με, α_T is the temperature coefficient expressed in MHz/Co, ΔT, is the temperature change which is the difference between the measured temperature and the ambient temperature, $\Delta\varepsilon$ is the strain change.

From this frequency shift, sensing is possible, but the issue here is that the shift is dependent on both strain and temperature. From the frequency shift alone, you cannot tell which change has occurred. To solve this issue, make the fiber optic cable strain-free by placing it near the object that is being monitored. Under this arrangement, strain becomes zero which eliminates the first term in Eq. (24).

Therefore, the Brillouin frequency shift is calculated as indicated in Eq. (25):

$$\nu_{B1} - \nu_{01} = \alpha_T \Delta T \tag{25}$$

ν_{B1} is the measured Brillouin frequency shift, ν_{01} is the reference Brillouin frequency shift at the ambient temperature, in MHz. **Figure 3** shows the linear relationship between the Brillouin frequency shift and the temperature change. The plot in the figure is based on an ambient temperature of 20°C, a reference Brillouin frequency shift of 16,650 MHz/C° and a temperature coefficient of 1.6 MHz/C°.

For strain measurement, it is recommended to use two fiber optic cables. One cable is attached to the object being monitored and another cable is made strain-free by placing it near the object being monitored.

The first cable which is attached to the object being monitored has these parameters: $\nu_{B1}, \nu_{01}, \alpha_\varepsilon, \Delta\varepsilon_1, \alpha_T, \Delta T_1$ and is used to monitor strain changes. The second cable, the strain-free cable, has these parameters: $\nu_{B2}, \nu_{02}, \alpha_\varepsilon, \Delta\varepsilon_2, \alpha_T, \Delta T_2$ and is used to monitor the temperature changes.

Assuming the change in temperature in the bonded fiber cable equals the change in the free fiber, then the strain can be calculated using Eq. (26):

$$\Delta\varepsilon_1 = \frac{(\nu_{B1} - \nu_{01}) - (\alpha_T \Delta T_2)}{\alpha_\varepsilon} \tag{26}$$

ν_{B1} and ν_{01} are the measured Brillion frequency shift and the reference Brillouin frequency shift for the first fiber optic cable, and ΔT_2 is the temperature change. The strain measurement is known as micro-strain ($\mu\mathcal{E}$). Stretching a fiber optic cable from 1 m to 1.000010 m results in a strain of ten micro-strains ($10\ \mu\mathcal{E}$).

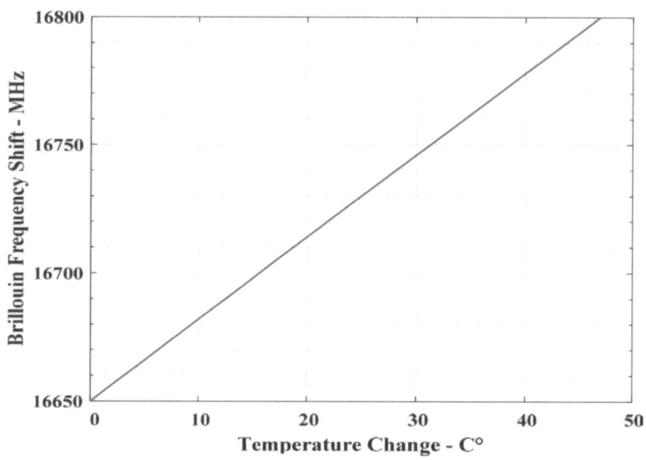

Figure 3.
Brillouin frequency shift – temperature change.

3.3 Location of the change

Calculating the time difference between the launched laser and the scattered laser, one can determine the location of temperature change or strain along the object being monitored using Eq. (27):

$$d = \frac{c\Delta t}{2n} \tag{27}$$

c is the speed of light in vacuum, Δt is the difference between the laser launch time and arrival time (scattered laser), and n is the refractive index of the fiber. As an example, let us assume that a fiber optic cable is placed in the vicinity of an onshore crude oil pipeline, and it is given that the time difference between the launched and scattered lasers is 25 μs. For a fiber optic cable having a refractive index of 1.5 and a speed of light in vacuum of 300,000 km/s, the temperature change may be detected at 2.5 km away.

3.4 The lowest detectable temperature and strain

The threshold which is the lowest detectable change is calculated as a function of minimum Brillouin frequency shift. From this value, a threshold can be established. Exceeding the threshold will indicate the occurrence of a change in temperature or strain. The minimum detectable Brillouin shift $\Delta\nu_B$ as a function of signal to noise ratio (SNR) is given by [2]:

$$\delta\nu_B = \frac{\Delta\nu_B}{\sqrt{2}(SNR)^{0.25}} \tag{28}$$

$\Delta\nu_B$ is the Brillouin spectral width and the SNR is signal to noise ratio. Using Eqs. (29) and (30) one can determine the minimum detectable changes in temperature and strain respectively.

$$\delta T = \frac{\Delta\nu_B}{\sqrt{2}\alpha_T(SNR)^{0.25}} \tag{29}$$

δT is the minimum detectable temperature change and α_T is the temperature coefficient expressed in MHz/C^o. As an example, let us assume that a system is designed to have a temperature coefficient = $1.52\ MHz/C^o$, SNR = 35 dB, and Brillouin spectral width = 25 MHz, the minimum detectable temperature change becomes: 1.55° C. The reader is advised to convert the SNR to a linear scale when using the Equation.

$$\delta\varepsilon = \frac{\Delta\nu_B}{\sqrt{2}\alpha_\varepsilon(SNR)^{0.25}} \tag{30}$$

$\delta\varepsilon$ is the minimum detectable strain change and α_ε *is* the strain coefficient expressed in $MHz/\mu\varepsilon$.

The spatial resolution is the minimum length of cable required to adequately detect a localized temperature change along the cable. Eq. (31) calculates the spatial resolution in terms of speed of light c in vacuum, the refractive index n of the fiber optic cable, and the pulse width τ.

$$\Delta z = \frac{c\tau}{2n} \tag{31}$$

For a pulse width of 10 ns, fiber optic cable refractive index of 1.5, at a speed of light 300,000 km/s in vacuum, the spatial resolution becomes 1 meter.

4. Summary and concluding remarks

This chapter presented a theoretical study on fiberoptic-based sensing. The stimulated Brillouin gain and loss, Brillouin spectrum, distributed Brillouin sensing techniques, the smallest observable changes, and the location of the changes were all covered in detail. Additionally, equations associated with light scattering behaviors, as well as equations associated with the characteristics mentioned above were derived and presented.

It was pointed out that a linear relationship exists between the frequency shift and the changes in temperature and strain. Furthermore, it was indicated that the signal to noise ratio and the Brillouin spectral width are the two determining factors of the smallest measurable variations in temperature and strain. The time difference between launched and scattered lasers can be used to determine the location of potential temperature and strain changes.

Future research should include experimental investigation of the behaviors of scattered light along the optical fiber in different environments, underwater, near shorelines, and areas with varying weather conditions. This will enable researchers to gain a better understanding and insight of the impact of scattered light on the optical fiber in different situations and environments.

Appendix A: Conversion of $\alpha_{dB/km}$ to $\alpha/_{\text{km}}$

As the signal propagates through the fiber, its power gets reduced due to the attenuation of the fiber optic cable. The output power through a fiber optic cable can be expressed in terms of the attenuation as [4]:

$$P_o = P_i e^{-\alpha L} \tag{32}$$

where P_i is the input power, α is the attenuation factor in km^{-1}, and L is the length of the fiber optic cable. α is usually provided with dB/km unit which requires a conversion to the appropriate km^{-1} unit through the following procedures. Recall that the equation for attenuation in dB is:

$$\alpha_{dB} = -10 \log_{10}\left(\frac{P_o}{P_i}\right) \tag{33}$$

To convert Eq. (33) to dB/km, simply divide the equation by L:

$$\alpha_{dB/km} = -\frac{10}{L} \log_{10}\left(\frac{P_o}{P_i}\right) \tag{34}$$

Rearrange Eq. (32) to solve for $\frac{P_o}{P_i}$:

$$\frac{P_o}{P_i} = e^{-\alpha L} \tag{35}$$

Substitute Eq. (35) into Eq. (34):

$$\alpha_{dB/Km} = -\frac{10}{L}\log_{10}\left(e^{-\alpha L}\right) \tag{36}$$

After simplifying, we end up with the following equation:

$$\alpha_{dB/km} = 10\left(\log_{10} e\right)\alpha = 4.343\alpha \tag{37}$$

Author details

Alireda Aljaroudi[1*] and Ahmed Aljaroudi[2]

1 Independent Engineering Research, Hamilton, ON, Canada

2 Research Collaborator, Hamilton, ON, Canada

*Address all correspondence to: aaa515@mun.ca

References

[1] Bao X, Chen L. Recent progress in distributed fiber optic sensors. Sensors. 2012;**2012**(12):8601-8639. DOI: 10.3390/s120708601. Available from: www.mdpi.com/journal/sensors

[2] Horiguchi T, Shimizu K, Kurashima T, Tateda M, Koyamada Y. Development of a distributed sensing technique using Brillouin scattering. Journal of Lightwave Technology. 1995; **13**:1296-1302

[3] Ruffin AB. Stimulated Brillouin scattering: An overview of measurements, system impairments, and applications. In: Symposium on Optical Fiber Measurements. National Institute of Standards and Technology – Special Publication 1024. 2004. pp. 23-28

[4] Agrawal GP. Nonlinear Fiber Optics. 3rd edition. San Diego, California, USA: Academic Press; 2001

[5] Bao X, Webb DJ, Jackson DA. 22-km distributed temperature sensor using Brillouin gain in an optical Fiber. Optics Letters. April 1, 1993;**18**(7)

[6] Smith JR. Characterization of the Brillouin loss spectrum for simultaneous distributed sensing of strain and temperature [MSc thesis]. Ottawa, Canada: Department of Physics, University of New Brunswick, Canada and National Library of Canada; 1999

[7] Diament P. Wave Transmission and Fiber Optics. New York: Macmillan; 1990

[8] Soto MA, Bolognini G, Pasquale FD. Long-range simplex-coded BOTDA sensor over 120km distance employing optical pre-amplification. Optics Letters. 2011;**36**(2):232-234

[9] Brown K. Improvement of a Brillouin scattering based distributed fiber optic sensor [PhD Thesis]. Ottawa, Canada: University of New Brunswick and Library & Archives; 2006

Chapter 3

Sensing Performance of Optical Waveguide

Veer Chandra

Abstract

Optical waveguide-based sensors are gaining popularity due to their cost-effectiveness, compact size, and high sensitivity. There are two primary techniques for designing these sensors: changes in effective refractive index or changes in the evanescent field ratio. The absorption property, specifically the evanescent field-based photonic sensing approach, is preferred over effective index-based sensing. This preference arises from the fact that the change in effective index is relatively smaller when the sensing material is present, compared to the change in the evanescent field. The absorption of light depends on the dimension and material of the sensor device, the operating wavelength, and the material being sensed. The percentage of evanescent field/light in the upper cladding/slot region of the photonic waveguide changes during its interaction with the molecules of the sensing material. Therefore, it is crucial to design photonic sensors with waveguides that have a relatively larger evanescent field in the upper cladding region. In slot waveguides, light is predominantly confined to the slot region, resulting in a higher evanescent field value. As the sensitivity of the photonic sensor depends on the percentage of evanescent field/light in the upper cladding region, the slot waveguide offers significantly higher sensitivity compared to other photonic waveguide structures.

Keywords: optical waveguide, slot photonic waveguide, evanescent field ratio, propagation loss, sensitivity, gas sensor

1. Introduction

Optical waveguide-based photonic sensors have gained significant popularity due to their advantageous characteristics, making them increasingly preferred in diverse applications. One key factor contributing to their popularity is their cost-effectiveness, making them an economically viable choice for various industries. The ultra-compact size of these sensors further enhances their appeal, allowing for integration into smaller devices and systems. One noteworthy attribute of optical waveguide-based photonic sensors is their high sensitivity. This heightened sensitivity makes them particularly suitable for applications where precise and accurate measurements are crucial. In fields such as biomedical research, chemical detection, and gas detection, where minute changes in parameters can be of great significance, these sensors offer a reliable and efficient solution. The miniaturization of sensors is a particularly intriguing aspect of this technology. Unlike traditional bulky sensors, optical waveguide-based photonic sensors can be seamlessly incorporated into a small chip area. This compact design not only facilitates easy integration into various devices but also opens up new possibilities for the development of smaller, more portable sensor systems. Another advantage lies in the ease of fabrication, owing to the compatibility of these sensors with complementary metal-oxide-semiconductor (CMOS) technology. The CMOS compatibility streamlines the manufacturing process, making it more efficient and cost-effective. This compatibility also allows for the integration of photonic sensors with existing electronic systems, creating synergies between optical and electronic components [1]. Designing a photonic sensor involves a keen observation of variations in either the effective refractive index or the evanescent field ratio. These parameters play a crucial role in the sensor's ability to detect changes in the surrounding environment or the material being sensed. There are two primary approaches for photonic sensing: effective index-based sensing and absorption property-based sensing. In the context of these approaches, the absorption property-based photonic sensing method is often favored. This preference stems from the recognition that, in the presence of the material or sample under consideration for sensing, the change in the effective refractive index is relatively smaller compared to the change in the evanescent field. The dominant factor contributing to this change in the evanescent field is the absorption of light by the sensing material. The evanescent field is an electromagnetic field that extends beyond the surface of the waveguide or sensor. Its variation is particularly sensitive to changes in the surrounding medium. When the sensing material interacts with this field, the absorbed light induces a noticeable alteration in the evanescent field, making it a reliable indicator of changes in the environment. This sensitivity to absorption phenomena allows for a more accurate and responsive detection mechanism. The magnitude of light absorption depends on several factors, including the dimensions and material composition of the sensor device, the operating wavelength of the light used, and the characteristics of the material to be sensed. The choice of these parameters influences the sensor's efficiency and specificity in capturing changes in the absorption properties of the surrounding medium [2, 3].

In the mid-infrared range, the authors of [4–6] have proposed the concept of an evanescent field-based photonic gas sensor employing strip, rib, and slot photonic waveguides. During its contact with the molecules of the sensing material, the photonic waveguide's upper cladding/slot region's percentage of light or evanescent field varies. As a result, the waveguides in the top cladding region of the photonic sensors must be built with a greater evanescent field. The slot waveguide, at the expense of a larger propagation loss [7–10], has the highest evanescent field ratio in the slot (upper cladding) region when compared to other photonic waveguide architectures, as the authors in [11, 12] have shown. Slot waveguides represent a distinctive type of waveguide design where light is primarily confined within a narrow slot region. This characteristic imparts unique advantages to slot waveguides, especially concerning their evanescent field properties. The evanescent field is a critical component in photonic sensors, and in the context of slot waveguides, its magnitude is notably higher compared to other waveguide structures [13–19]. The confinement of light within the slot region is a defining feature of slot waveguides. This results in a more significant evanescent field, which is the electromagnetic field extending beyond the core of the waveguide. In practical terms, the higher evanescent field in slot waveguides means that a larger portion of the light is near the waveguide's surface, making it more responsive to changes in the surrounding environment. However, it is essential to acknowledge that slot waveguides may incur propagation losses. To address these losses, a compensatory strategy involves using a light source with relatively high input power. This higher power input helps counterbalance the losses experienced during the propagation of light through the waveguide. While this compensatory measure may require more power, it contributes to maintaining the overall efficiency of the photonic sensor system. The significance of the evanescent field in photonic sensors cannot be overstated. The sensitivity of these sensors relies heavily on the percentage of evanescent field or light present in the upper cladding region, where interactions with the external environment occur. In the case of slot waveguides, the inherently higher evanescent field values translate into a sensor with significantly heightened sensitivity.

In the contemporary landscape, Micro-Electro-Mechanical System-based sensors have gained widespread acceptance as essential electronic devices [20–24]. However, a promising counterpart, known as the Micro-Opto Mechanical System, has emerged as a compelling technology [1, 25, 26]. This amalgamation of features positions the Micro-Opto Mechanical System as a technology with immense potential, raising expectations that it may play a dominant role in the global sensor market in the near future. The distinguishing feature of the Micro-Opto Mechanical System lies in its ability to merge photonics and micro-machining. This convergence results in a sensor system that harnesses the benefits of both domains. High sensitivity is a particularly noteworthy advantage of the Micro-Opto Mechanical System, making it adept at capturing subtle changes in the environment or the parameters being measured. Additionally, the ease of fabrication associated with the Micro-Opto Mechanical System facilitates cost-effective manufacturing processes, contributing to its attractiveness for widespread adoption. The synergy between the Micro-Opto Mechanical System and CMOS enhances the overall efficiency and functionality of the sensor devices, making them more versatile and adaptable to various applications. Anticipating the future trajectory, it is plausible to envision Micro-Opto Mechanical System-based sensor devices occupying a substantial share of the global sensor market. This projection is rooted in the Micro-Opto Mechanical System's inherent advantages, including high sensitivity, ease of fabrication, and compatibility with CMOS technology. Moreover, beyond the capacitive sensing approach commonly associated with Micro-Electro-Mechanical Systems, the literature showcases various photonic sensors based on optical interference methods [27–29]. These sensors leverage optical interference phenomena to precisely measure deflection, providing accurate information about the exerted pressure. This diversified approach in photonic sensing methods contributes to the expanding capabilities of Micro-Opto Mechanical System -based sensors, further solidifying their potential dominance in the sensor market.

2. Working principles of optical waveguide-based sensor

The sub-wavelength dimensions of dielectric waveguides contribute to a fascinating phenomenon in the form of a small fraction of electromagnetic (EM) modes that extend beyond the waveguide itself. This extension of electromagnetic fields beyond the waveguide is referred to as the evanescent field. The significance of this evanescent field, quantified by the evanescent field ratio, becomes pivotal in understanding the behavior of light propagation along the waveguide. Several factors influence the evanescent field, including the cross-sectional geometry of the waveguide in relation to the wavelength of the propagating light and the optical properties of the materials involved [30]. The evanescent field, because of its sub-wavelength nature, plays a crucial role in applications such as gas sensing. Gas sensing is particularly demanding due to the typically low absorption coefficients associated with most gases. In the context of gas sensing with dielectric waveguides, achieving a high evanescent field ratio becomes essential. A high evanescent field ratio indicates that a substantial portion of the electromagnetic modes propagating along the waveguide extend outside its physical boundaries. The importance of a high evanescent field ratio in gas sensing is closely tied to the fact that the intrinsic damping of the evanescent field should be kept low. In other words, the evanescent field should be capable of penetrating into the surrounding medium, and its energy should not be rapidly attenuated. This characteristic is crucial for effective gas sensing because when an absorbing medium such as a gas is introduced in the vicinity of the waveguide, the evanescent field is absorbed by the gas. This absorption attenuates the guided wave within the waveguide, leading to a reduction in intensity at the output port of the waveguide [31].

2.1 Design and modeling

The sensor test structures can be considered as silicon slot waveguides of approximately 1 cm in length, specifically engineered for the purpose of gas sensing. To enhance their functionality, taper structures and launch pads have been incorporated at both ends of the waveguides. Notably, a grating coupler is positioned on the launch pad to facilitate the coupling of mid-infrared radiation (MIR) from an external MIR source into the waveguide. Additionally, the grating coupler serves the purpose of efficiently extracting the light from the waveguide, directing it toward a detector for further analysis. The detailed design of this grating coupler, employing two-dimensional finite element models, has been elucidated in [32]. The schematic representation in **Figure 1** provides an overview of the measurement configuration employed. It includes a gas cell housing the test chip with the silicon waveguide structures. The gas cell is an integral component of the experimental setup, allowing controlled exposure of the sensor to different gas environments for testing purposes. The schematic captures the essential elements of the setup, depicting the waveguide, taper structures, launch pads, and the grating coupler. The gas cell within the configuration serves as a controlled environment where the gas sensing experiments are

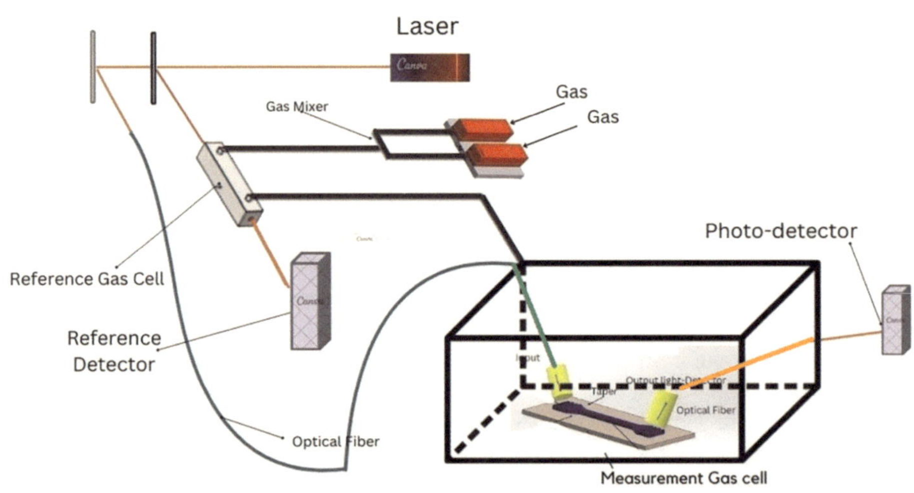

Figure 1.
Schematic representation of the measurement setup.

conducted. The integration of the grating coupler allows for the efficient interaction of mid-infrared radiation with the waveguide, enabling precise and selective sensing. The out-coupling of light onto a detector facilitates the measurement of the guided wave, providing valuable data related to the absorption characteristics of gas in the mid-infrared spectrum.

2.2 Absorption-based sensing

Sensing mechanisms relying on the absorption properties of materials are generally favored over those based on the effective refractive index. This preference arises from the observation that changes in the effective refractive index have a relatively minor impact compared to alterations in the evanescent field, primarily induced by absorption phenomena. In absorption-based sensing, the evanescent field engages with the molecules of the sensing element, leading to the absorption of a certain amount of power by these molecules, depending on the wavelength. Consequently, designing photonic sensors involves the use of waveguides with larger evanescent fields, thereby ensuring higher sensitivity, particularly in the near-IR region. The sensitivity can be quantified using Eq. (1) [8, 12].

$$Sensitivity(s) = -\eta EL\ exp\left(-\eta ECL - \alpha_{wvg}L\right) \tag{1}$$

where, η, E, C, α_{wvg}, and L are the evanescent field fraction, absorption coefficient, concentration, intrinsic waveguide loss, and length of the waveguide respectively. The evanescent field fraction is defined as the amount of power in the upper etched cladding region, interacting with the sensing environment.

2.3 Evanescent field ratio

The evanescent field ratio holds significant importance in the development of devices and sensors based on evanescent field absorption. This parameter is defined as the proportion of power in the upper cladding region, engaged with the environment or material intended for sensing, relative to the total incident power on the device. The evanescent field ratio is predominantly influenced by the geometric characteristics of the photonic waveguide and operating wavelength. The evanescent field ratio can be expressed by Eq. (2) below, highlighting its dependence on specific aspects of the waveguide's design and the wavelength employed.

$$\eta = \frac{\iint_{evan} \vec{S}.\vec{n}\,dx\,dy}{\iint_{Total} \vec{S}.\vec{n}\,dx\,dy} \tag{2}$$

where, η is the evanescent field ratio, n is the normal vector to the waveguide cross-section and S is the Pointing vector of the mode field in the waveguide.

2.4 Propagation loss

In addition to the evanescent field ratio, the analysis of sensor performance also hinges on the propagation loss within the waveguide. The propagation loss is a crucial factor influenced by the geometrical characteristics of the waveguide and is particularly dependent on the imaginary part of the effective refractive index. This relationship is mathematically expressed in Eq. (3) below [27–29]. Therefore, understanding and controlling waveguide geometries and the associated imaginary component of the effective refractive index is pivotal in comprehending and optimizing the performance of sensor designs.

$$Loss\ (dB) = 10 \times log_{10}(e) \times \frac{4\pi \times Im(n_{eff})}{\lambda} \tag{3}$$

where, $Im(n_{eff})$ is the imaginary part of the effective refractive index, and λ is the operating wavelength. From Eq. (3), it is clear that for a fixed wavelength, the propagation loss varies only with $Im(n_{eff})$. The slot waveguide usually offers higher evanescent field ratio values at the cost of higher propagation loss, which can be overwhelmed by the enhanced incident power.

2.5 Optimum length of waveguide

Moreover, the optimum length (L_{opt}), which can be found for a photonic sensor or device by solving the equation $\frac{ds}{dL} = 0$, from Eq. (1) above, is the waveguide length at which the sensitivity is maximal and beyond that the sensitivity starts to erode due to waveguide loss. Consequently, it can be stated as shown in the following Eq. (4),

$$L_{opt} = \frac{1}{\eta \varepsilon C + \alpha_{wvg} L} \tag{4}$$

3. Design and sensing performance of optical waveguide

3.1 Waveguide design

Normally, three distinct types of waveguides—namely strip, rib, and slot waveguides—can be designed for the purpose of absorption-based gas sensing in the mid-infrared (mid-IR) region, as illustrated in **Figure 2**. The waveguides are crucial components in the context of detecting gases, particularly emphasizing their application in the absorption of infrared radiation. The parameters width and height of the waveguide, respectively, provide key dimensions for the characterization of the waveguide structure. In the case of rib waveguides, an additional parameter rib layer, is introduced, representing the thickness of the dielectric slab layer. For slot waveguides, the slot gap between two high-index arms is a crucial parameter. These specifications are fundamental in tailoring the waveguide geometry to optimize its performance for gas sensing applications [12, 33, 34]. As a practical example, the chapter delves into the design of a specific waveguide intended for the absorption line of CH_4 at a wavelength of 3.31 μm in the mid-IR range. This wavelength selection aligns with the characteristic absorption behavior of CH_4 molecules in the mid-IR spectrum. By strategically configuring the dimensions and structure of the waveguide, the aim is to enhance its effectiveness in facilitating the absorption of infrared radiation, thereby enabling precise and sensitive gas sensing capabilities. This design approach emphasizes the importance of customization and optimization of waveguide parameters to suit the specific absorption characteristics of the target gas, showcasing the

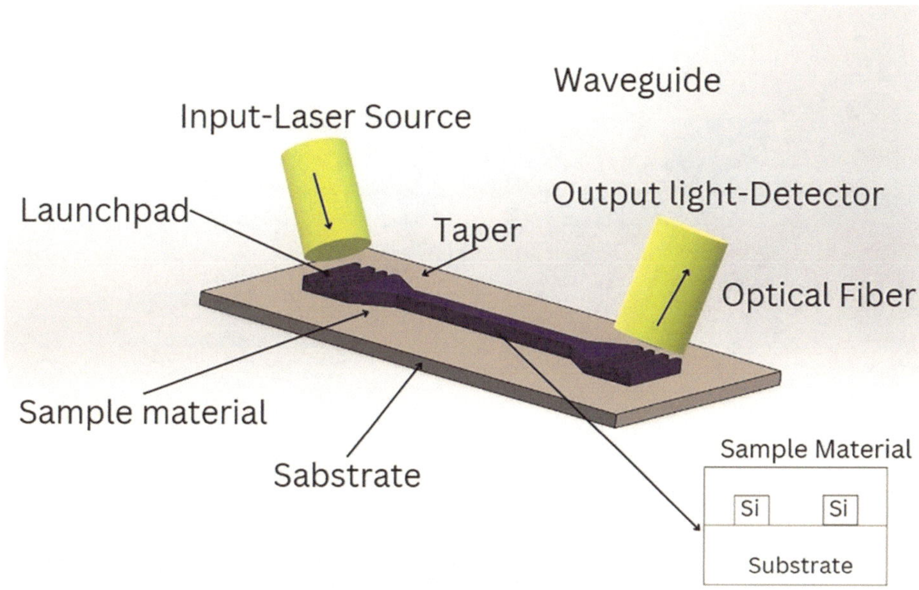

Figure 2.
The measurement setup along with waveguide configuration and optical fibers, with input and output light.

versatility and adaptability of waveguides for mid-IR gas sensing applications. The simulated η values will be used in the case study of CH_4 sensing in the next section. Since the gas in a slot waveguide will interact with the evanescent field at the waveguide's two sidewalls as well as on top, both quasi-TE and quasi-TM modes are interesting for field augmentation. To meet the electrical displacement continuity requirement, the dominating electrical field for the quasi-TE mode is in the horizontal direction, and it will be strengthened at the gas interface with the sidewalls (**Figures 3–5**).

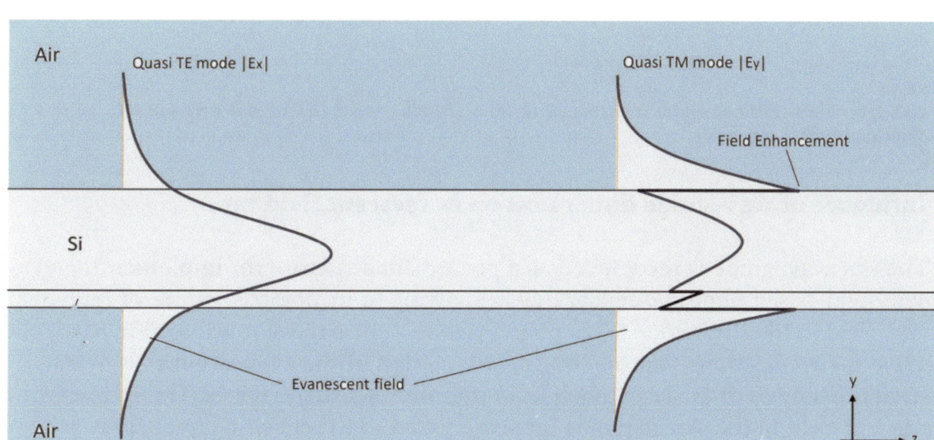

Figure 3.
Diagram displaying a dielectric waveguide's side view, illustrating the electric field distribution of the fundamental quasi-TE and quasi-TM modes.

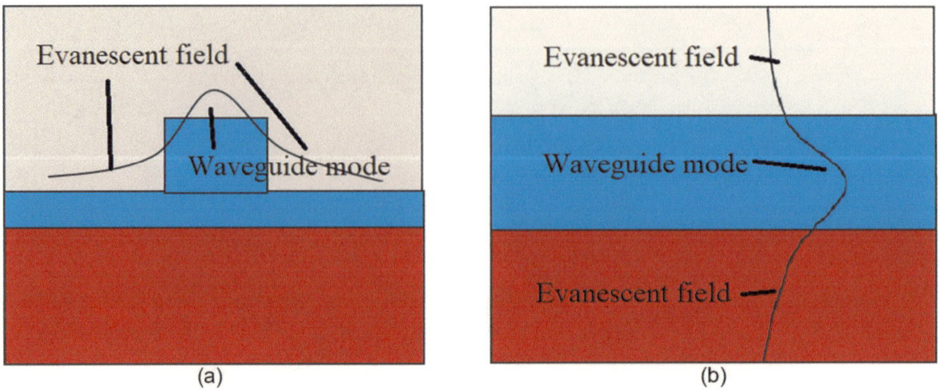

Figure 4.
Schematics of the evanescent absorption gas sensor: (a) cross-sectional view and (b) side view.

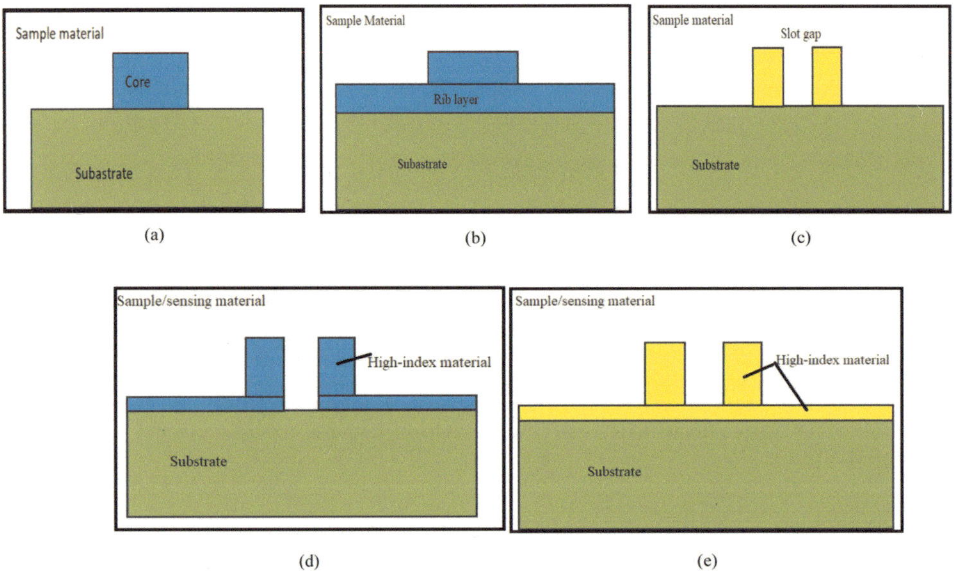

Figure 5.
Different types of possible waveguide structures: (a) strip, (b) rib (c) slot (d) partial-strip-loaded slot, and (e) full-strip-loaded slot waveguide.

3.2 Influence of waveguide dimension on evanescent field ratio

The slot waveguide is recognized as a prominent structure for implementing evanescent field-based photonic sensor devices, owing to its notable feature of high evanescent field in the slot region. This phenomenon makes the slot waveguide particularly effective for applications that rely on the interaction of the evanescent field with external substances or analytes, such as in photonic sensing devices. The efficacy of the evanescent field in the slot region is intricately tied to the specific dimensions and structure of the waveguide. The amount of evanescent field present in the slot region is a critical factor that directly influences the performance and sensitivity of the sensor. Therefore, it becomes imperative to carefully examine and understand the impact of variations in the waveguide's dimensions on the evanescent field ratio for all three

considered slot waveguide structures. The term evanescent field ratio refers to the concentration of the electric field within the waveguide structure, particularly in the slot region. This distribution is crucial for enhancing the interaction between the waveguide and the external environment, facilitating improved sensitivity in sensing applications. By visualizing and analyzing the effects of changes in waveguide dimensions on the electric field distribution, researchers and practitioners gain insights into how alterations in the structural parameters influence the sensor's performance.
This understanding is pivotal for optimizing the design of slot waveguides to achieve desired sensing characteristics and to tailor the sensor's response to specific applications. Hence, the chapter underscores the importance of investigating the intricate relationship between waveguide dimensions and evanescent field distribution to advance the development of efficient and reliable evanescent field-based photonic sensor devices.

In the investigation of conventional slot waveguides, the parameter 'W' has been systematically varied within the range of 450–650 nm for two distinct slot gaps (G) of 80 and 120 nm, as illustrated in **Figure 6a**. The graphical representation indicates a noticeable decrease in the percentage of evanescent field ratio with the initial increase in the width of the waveguide. Remarkably, for a slot gap of 80 nm, a sudden surge in evanescent field ratio can be observed at a waveguide width of 510 nm, reaching its maximum value (0.3) around the width of 520 nm. Prior to the width of 510 nm, light confinement predominantly occurs in the side arms (Ge) or lower-cladding regions, whereas after the width of 520 nm, light propagation primarily takes place through the slot region. Beyond the width of 520 nm, the confinement of light in the slot

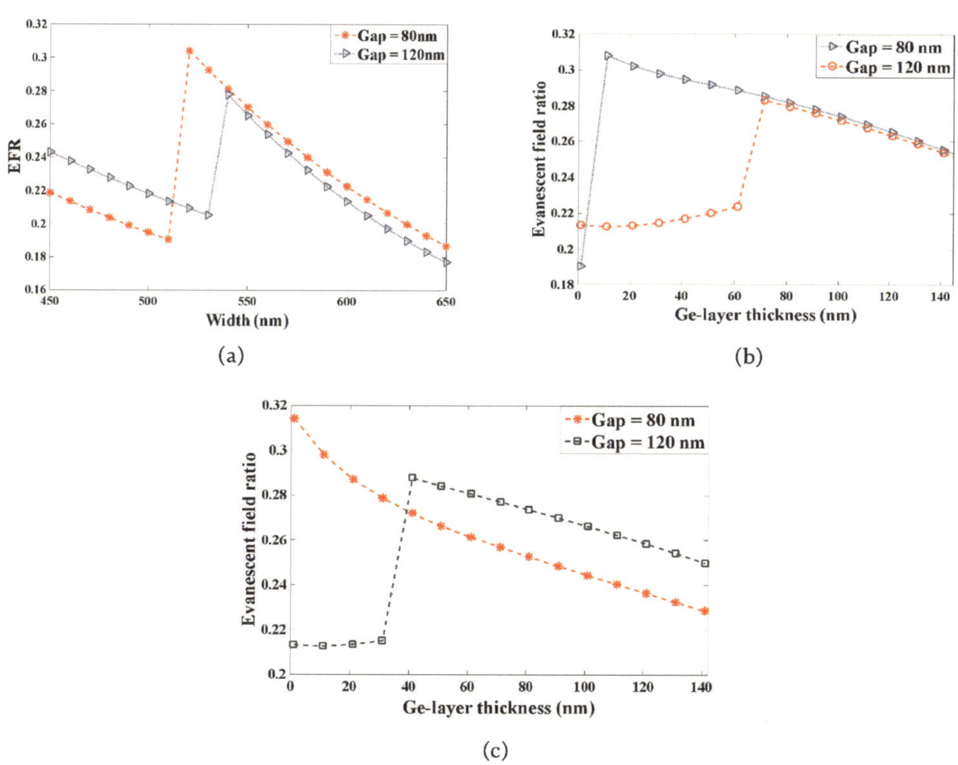

Figure 6.
Evanescent field ratio variations at two distinct slot gaps with thickness in (a) typical slot waveguide, (b) partial-strip-loaded slot waveguide, and (c) full-strip-loaded slot waveguide [35].

region gradually diminishes with increasing arm widths. A similar trend is identified for a slot gap of 120 nm, with light confinement in the slot region initiating at the width of 530 nm, peaking at width of 540 nm, before gradually decreasing. Extensive simulation analysis reveals that achieving light confinement in the slot region is challenging for lower values of slot gap, potentially resulting in light leakage into the upper/lower-cladding and high-index regions. Notably, substantial light propagation is observed for a slot gap value around 80 nm, leading to significant evanescent field values. In the context of evanescent field analysis for different slot waveguide structures, the thickness ('t') of the germanium (Ge) layer is varied up to 140 nm [35]. This exploration aims to identify the optimal Ge-layer thickness that maximizes the evanescent field ratio in the slot region. The analysis considers two slot gaps of 80 and 120 nm, presented in **Figure 6b** and **c**, respectively, for two different types of slot waveguides i.e., partial-strip-loaded and full-strip-loaded. Notably, similar trends in the graph have been reported in recent literature, emphasizing the variations in the confinement factor in slot waveguides. Other studies by different authors have also showcased analogous transitions in their results concerning the confinement factor of slot waveguides [36–38].

Building upon the insights gained from the analysis of conventional slot waveguides and the meticulous simulations involving partial-strip-loaded and full-strip-loaded slot waveguides, the study focuses on a fixed waveguide width value of 510 nm. In **Figure 6b**, dedicated to the partial-strip-loaded slot waveguide, the maximum levels of the evanescent field are observed at approximately 30.22 and 28.31% for slot gaps of 80 and 120 nm, respectively. These peaks occur at specific thickness values, namely 10 and 60 nm. Similarly, **Figure 6c**, dedicated to the full-strip-loaded slot waveguide, presents the variations in evanescent field ratio concerning thickness. Maximum evanescent field ratio values of about 31.4 and 28.77% are noted for the same respective slot gaps, accompanied by corresponding thickness values of 0 and 40 nm. Interestingly, following the attainment of the highest evanescent field values in both different types of slot waveguides namely partial-strip-loaded and full-strip-loaded, an increase in thickness leads to a subsequent decrease in the evanescent field. It is noteworthy that the identified values of width and thickness for the considered slot waveguides align with the configurations that yield the maximum evanescent field in the slot region. However, beyond just the evanescent field, the propagation loss plays a pivotal role in determining the optimal dimensions of waveguides. Striking a balance between achieving higher sensitivity and minimizing propagation loss is crucial, and these considerations are further delved into in the subsequent subsections below.

3.3 Influence of waveguide dimension on propagation loss

Propagation loss is intricately linked to the choice of materials and the geometric characteristics of photonic waveguides. Building upon the preceding analysis, there is an expectation that depositing a Ge-layer on CaF_2 may lead to a reduction in propagation loss in slot waveguide structures. To explore the impact of variations in waveguide dimensions on propagation loss, the width of the slot waveguide is once again systematically adjusted, ranging from 450 to 650 nm, for slot gaps (G) of 80 and 120 nm. Through simulation analysis, the imaginary part of the effective refractive index can be derived. Subsequently, using these imaginary values of effective refractive indices values for different slot waveguide dimensions in Eq. (3), the propagation loss in slot waveguides can be determined. In **Figure 7a**, the relationship between propagation loss and width of conventional slot waveguide is illustrated. As the width

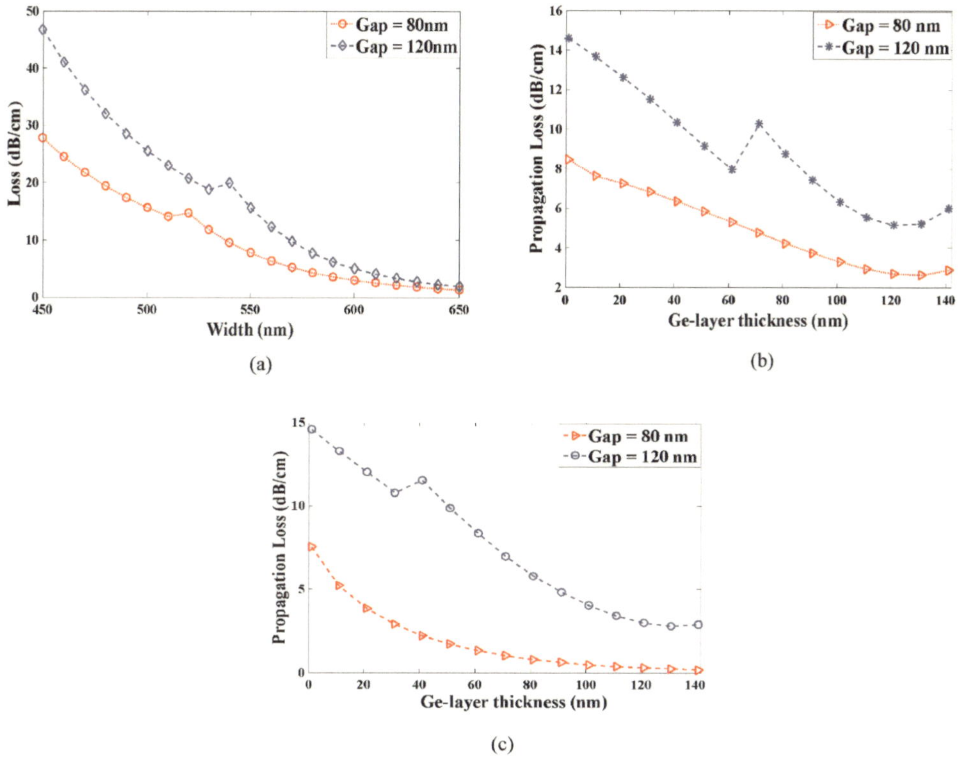

Figure 7.
Propagation loss variations versus thickness at slot gaps of 80 nm and 120 nm for (a) conventional slot waveguide, (b) partial-strip-loaded slot waveguide, and (c) full-strip-loaded slot waveguide [35].

increases, there is a noticeable decrease in propagation loss. However, as highlighted in the preceding subsection, for slot gap of 80 nm, substantial light confinement in the slot region commences at a waveguide width of 510 nm, reaching its maximum at a waveguide width of 520 nm. Consequently, the propagation loss begins to increase at waveguide width 510 nm, peaking at a waveguide width 520 nm, before experiencing a continuous decrease. This suggests that in slot waveguides, achieving high light evanescent field comes at the expense of elevated propagation loss. Moreover, to strike a balance between low propagation loss and significant light confinement, the choice of waveguide width becomes crucial. For instance, at a waveguide width of 550 nm with a slot gap of 80 nm, the evanescent field decreases by only 3%, accompanied by a substantial reduction in propagation loss, nearly 7 dB/cm compared to that at waveguide width of 520 nm. Similarly, for a slot gap of 120 nm, selecting a waveguide width of 560 nm results in a reduction in the evanescent field and corresponding propagation loss by around 2.4% and 7.6 dB/cm, respectively, when compared to a waveguide width of 540 nm. This underscores the importance of judiciously selecting the waveguide width to achieve optimal trade-offs between propagation loss and light confinement, particularly in the context of varying slot gaps.

Furthermore, in the analysis of propagation losses within different slot structures, i.e., partial-strip-loaded and full-strip-loaded, a fixed waveguide width value of 510 nm is maintained, while the thickness of the waveguide layer can be systematically varied up to 140 nm. This examination is conducted for slot gaps (G) of 80 and

120 nm, as illustrated in **Figure 7b** and **c**, respectively. The results clearly indicate that for a slot gap of 80 nm, the propagation loss consistently decreases with the incremental rise in the thickness layer, exhibiting a smooth trend without any abrupt changes in propagation loss observed at intermediate thickness layer values, as noted in the case of a slot gap of 120 nm. Similar to the earlier analysis, the careful selection of waveguide bottom layer thickness becomes paramount to achieving a significant evanescent field with low propagation loss. **Figure 7** highlights that both waveguide width and thickness values contribute to the reduction in propagation loss, with substantially lower propagation loss apparent for smaller values of slot gap. Furthermore, in terms of propagation loss, the full-strip-loaded type structure outperforms the partial-strip-loaded and notably surpasses the conventional slot waveguide. This suggests that both different slot waveguide structures offer viable options for the design of gas sensors based on the absorption property of sample material in photonic waveguide, emphasizing their potential suitability and superior performance in such applications.

3.4 Sensitivity analysis for methane gas

To enhance the sensitivity of photonic waveguides, it is imperative to maximize the evanescent field while minimizing propagation loss. However, in typical slot waveguides, there exists a trade-off between achieving a higher evanescent field and

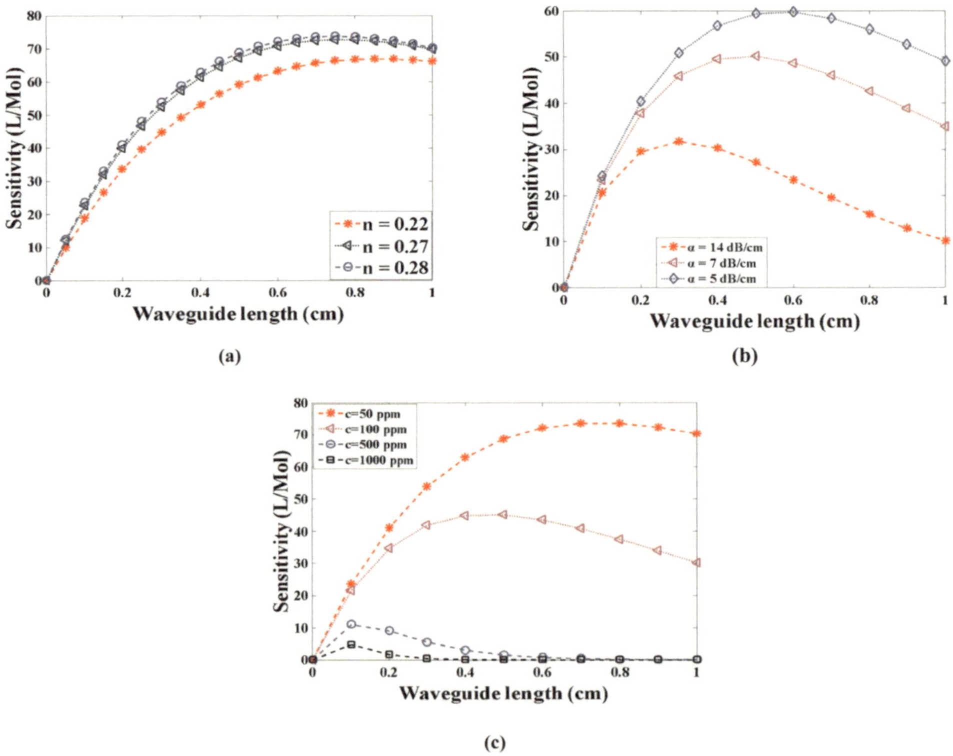

Figure 8.
Sensitivity as a function of waveguide lengths for the different waveguide structures: (a) at a fixed propagation loss of 3 dB/cm and slot gap of 80 nm, (b) at a fixed evanescent field ratio of 0.28 and (c) four distinct methane gas concentrations for a full-strip-loaded slot waveguide at slot gap of 80 nm and [35].

keeping propagation loss low. Striking a balance between these two factors is crucial to attain a satisfactory level of sensitivity. Utilizing Eq. (3), sensitivity values for all three slot waveguide structures can be estimated, considering an absorption coefficient of 960.96 L/Mol-cm [35] to quantify device sensitivity. **Figure 8a–c** portray the variations in sensing capabilities concerning waveguide lengths for different evanescent field ratio values in the slot region, propagation losses, and gas concentrations across all different slot waveguide structures. Specifically, for a slot gap of 80 nm in **Figure 8a**, the sensitivity variations are illustrated for three distinct evanescent field ratios (0.22, 0.27, and 0.28) obtained for the different slot waveguide structures, respectively. These analyses are conducted while maintaining a fixed propagation loss of 3 dB/cm.

The figure distinctly illustrates that among the three different structures of slot waveguide, the full-strip-loaded slot waveguide exhibits the highest sensing ability. However, the sensitivities of both partial-strip-loaded and full-strip-loaded slot waveguides are closely aligned, given their similar evanescent field values at a fixed propagation loss of 3 dB/cm. Furthermore, the influence of varying waveguide lengths can also be discussed in **Figure 8b**, considering different propagation losses (approximately 14, 7, and 5 dB/cm) achieved for different slot waveguide structures. In this analysis, a fixed evanescent field ratio of 0.30 and slot gap of 80 nm has been maintained. Notably, while maintaining identical evanescent field values for all slot waveguide structures, their propagation losses differ due to distinct design structures. Additionally, the partial-strip-loaded and conventional slot waveguides exhibit reduced sensing abilities by 5–15 and 20–40 L/Mol, respectively, compared to the full-strip-loaded slot waveguide. Consequently, from **Figure 8a** and **b**, it is evident that sensitivity increases with rising propagation loss and decreasing evanescent field value, underscoring the superior performance of the full-strip-loaded slot waveguide. Similarly, **Figure 8c** analyses the sensitivity variations in terms of waveguide lengths for a full-strip-loaded slot waveguide at different concentrations of methane gas (50, 100, 500, and 1000 ppm). The figure reveals that sensitivity decreases with an increase in gas concentrations. This phenomenon is attributed to the heightened interaction of gas molecules with the evanescent field as the concentration rises, leading to increased light/evanescent field absorption and subsequently elevated losses, thereby decreasing sensitivity. However, for a specific gas concentration, sensitivity initially rises, reaches its peak at an optimum length, and then starts to decline beyond this optimal length.

3.5 Comparison for optical detection of different gas

The primary emphasis of the current research lies in the development of the sensing element dedicated to a light source with a specific wavelength of 3.31 μm. However, it is crucial to acknowledge that the effectiveness of photonic sensors is not solely dependent on the sensing element; the photodetector also plays a pivotal role in the overall design and analysis of such sensors. In the context of methane gas sensing devices, the focus shifts toward the implementation of a PbTe type of photodetector. This photodetector is identified as a suitable choice for detecting light with a wavelength of 3.31 μm emanating from the sensor device. The unique advantage of PbTe photodetectors lies in their capability for monolithic integration, meaning they can be seamlessly incorporated into the sensor's structure. The integration of PbTe photodetectors contributes to the reliability of the sensing system, particularly in the detection of mid-infrared (mid-IR) light waves. This reliability is vital for applications like

Gas/sample	Operating wavelength (μm)	Detection limit (ppm)	Detector	Ref.
CO_2	4.26	400	—	[40]
CO_2	4.26	—	Photodiode	[41]
CO_2	4.26	30	Bolometer	[42]
CH_2O	0.34	4.3	Photodiode	[43]
CH_4	1.65	11		[44]
CH_4	3.31	10.3	Pyroelectric	[45]
CO_2	4.26	50	Thermopile	[46]

Table 1.
List of sample gas and its operating wavelength along with detector.

methane gas sensing, where accuracy and precision in detecting specific wavelengths are essential. Therefore, by focusing on both the sensing element and the choice of a suitable photodetector, the research aims to enhance the overall performance and reliability of photonic sensors, particularly in the targeted detection of 3.31 μm wavelength light in methane gas sensing applications [39].

Optical detectors for gas sensing, aimed at measuring I(l), have demonstrated successful implementation across a spectrum of gases including acetone (C_3H_6O), ammonia (NH_3), carbon dioxide (CO_2), formaldehyde (CH_2O), nitric oxide (NO), carbon monoxide (CO), methane (CH_4), and methanol (CH_3OH). Diverse methodologies have been employed to construct gas sensors utilizing optical detection, as outlined in **Table 1**. These approaches encompass designs featuring tube-like gas cells established between face-to-face configured emitters and detectors, dome-like gas cells with planar configured emitters and detectors, open cells, cavity-enhanced cells, and waveguides utilizing evanescent-field interaction. The selection of light sources for these gas sensors has been varied, encompassing Micro-Electro Mechanical System heaters, LEDs, distributed feedback lasers (DFBs), and quantum cascade lasers (QCLs). Concurrently, an array of detectors, including photodiodes, thermopiles, pyroelectric detectors, and photoconductive detectors, has been utilized in these optical gas sensing configurations. The versatility in the design and components employed underscores the adaptability of optical gas sensors across a wide range of gases and applications.

4. Conclusion

The sensing performance of optical waveguides refers to the ability of these waveguides to effectively and accurately detect changes or variations in the surrounding environment, typically with respect to optical properties. Optical waveguides are structures designed to confine and guide light, enabling the transmission of optical signals with minimal loss. Evaluating the sensing performance involves assessing how well these waveguides can capture, transmit, and respond to optical signals in the context of specific applications, such as sensing various physical or chemical parameters. Key factors contributing to the sensing performance of optical waveguides include sensitivity. Sensitivity refers to the waveguide's ability to detect small changes in the surrounding environment, often expressed as the minimum detectable change in a parameter of interest. Moreover, the effectiveness of optical waveguides in sensing applications is influenced by their design, materials, and the specific detection mechanisms employed. Waveguides can be engineered to interact with external factors such as, pressure or chemical concentrations, allowing for a diverse range of sensing applications. Different types of waveguides, including strip and slot photonic waveguides, may exhibit variations in their sensing capabilities based on their inherent characteristics. The sensing performance of optical waveguides encompasses their ability to sensitively, accurately, and promptly detect changes in the optical environment. This property makes them valuable components in a wide array of sensing applications, ranging from environmental monitoring to biomedical diagnostics and gas sensing analysis.

Acknowledgements

The authors gratefully acknowledge DST-FIST, SRM Institute of Science and Technology Kattankulathur for their support.

Declarations

Ethics approval: The authors ensure that accepted principles of ethical and professional conduct have been followed during this research work.

The authors of this manuscript declare no conflicts of interest. This research does not involve any Human Participants and/or Animals.

Consent to Participate: Consent was obtained from the author who contributed to the research work.

Consent for Publication: The authors give full consent for publication of this research work.

Conflict of Interest: The author declares no conflicts of interest.

Funding: There is no funding involved in this research work.

Author details

Veer Chandra
Department of Electronics and Communication Engineering, SRM Institute of Science and Technology Kattankulathur, India

*Address all correspondence to: veerc@srmist.edu.in

References

[1] Bogaerts W, Baets R, Dumon P, Wiaux V, Beckx S, Taillaert D, et al. Nanophotonic waveguides in silicon-on-insulator fabricated with cmos technology. Journal of Lightwave Technology. 2005;**23**(1):401

[2] Butt MA, Degtyarev SA, Khonina SN, Kazanskiy NL. An evanescent field absorption gas sensor at mid-IR 3.39 μm wavelength. Journal of Modern Optics. 2017;**64**(18):1892-1897

[3] Ranacher C, Consani C, Vollert N, Tortschanoff A, Bergmeister M, Grille T, et al. Characterization of evanescent field gas sensor structures based on silicon photonics. IEEE Photonics Journal. 2018;**10**(5):2700614

[4] Ranacher C, Consani C, Tortschanoffa A, Jannesarib R, Bergmeisterc Grillec MT, Jakoby B. Mid-infrared absorption gas sensing using a silicon strip waveguide. Sensors and Actuators A: Physical. 2018;**277**:117-123

[5] Ranachera C, Tortschanoffa A, Moridi C, Grille M, Jakoby T, Bernhard. Photonic gas sensor using a silicon strip waveguide. Proceedings. 2017;**1**(547):1-5

[6] Korsakova SV, Romanova EA, Velmuzhov AP, Kotereva TV, Sukhanov MV, Shiryaev VS. Analysis of characteristics of the sensing elements for the fiber-based evanescent wave spectroscopy. Optics and Spectroscopy. 2018;**125**(3):416-425

[7] Ding R, Baehr-Jones T, Kim WJ, Xiong X, Bojko R, Fedeli JM. Low-loss strip-loaded slot waveguides in silicon-on-insulator. Optics Express. 2010;**18**: 25061-25067

[8] Almeida VR, Xu Q, Barrios CA, Lipson. Guiding and confining light in void nanostructure. Optics Letters. 2004;**29**:1209-1211

[9] Debnath K, Khokhar AZ, Boden SA, Arimoto H, Oo SZ, Chong HM, et al. Low-loss slot waveguides with silicon (111) surfaces realized using anisotropic wet etching. Frontiers in Materials. 2016; **3**:1-5

[10] Rifat AA, Ahmed R, Bhowmik BB. SOI Waveguide-Based Biochemical Sensors. In: Hameed M, Obayya S. editors. Computational Photonic Sensors. Cham: Springer; 2019. DOI: 10.1007/978-3-319-76556-3_16

[11] Butt MA, Khonina SN, Kazanskiy NL. Silicon on silicon dioxide slot waveguide evanescent field gas absorption sensor. Journal of Modern Optics. 2017;**65**(2):174-178

[12] Huang Y, Kalyoncu SK, Zhao Q, Torun R, Boyraz O. Silicon-on-sapphire waveguides design for mid-IR evanescent field absorption gas sensors. Optics Communications. 2014;**313**: 186-194

[13] Soref RA, Emelett SJ, Buchwald WR. Silicon waveguided components for the long-wave infrared region. Journal of Optics A: Pure and Applied Optics. 2006; **8**(10):840

[14] Tien E, Huang Y, Gao S, Song Q, Qian F, Kalyoncu SK, et al. Discrete parametric band conversion in silicon for mid-infrared applications. Optics Express. 2010;**18**:21981-21989

[15] Baehr-Jones T, Spott A, Ilic R, Spott A, Penkov B, Asher W, et al. Silicon-on-sapphire integrated waveguides for the mid-infrared. Optics Express. 2010;**18**:12127-12135

[16] Li F, Jackson SD, Grillet C, Magi E, Hudson D, Madden SJ, et al. Low propagation loss silicon-on-sapphire waveguides for the mid-infrared. Optics Express. 2011;**19**:15212-15220

[17] Wang Z, Liu H, Huang N, Sun Q, Wen J, Li X. Influence of three-photon absorption on mid-infrared cross-phase modulation in silicon-on-sapphire waveguides. Optics Express. 2013;**21**: 1840-1848

[18] Schmitt K, Oehse K, Sulz G, Hoffmann C. Evanescent field sensors based on tantalum pentoxide waveguides—A review. Sensors. 2008; **8**(2):711-738

[19] Parriaux O, Veldhuis GJ. Normalized analysis for the sensitivity optimization of integrated optical evanescent-wave sensors. Journal of Lightwave Technology. 1998;**16**(4):573-582

[20] Zinoviev K, Dominguez C, Plaza JA, Busto VJC, Lechuga LM. A novel optical waveguide microcantilever sensor for the detection of nanomechanical forces. Journal of Lightwave Technology. 2006; **24**(5):2132-2138

[21] Kumar BK, Basavaprabhu S. A critical review of MEMS capacitive pressure sensors. Sensors and Transducers. 2015;**187**:120-128

[22] Chandra M, Chen S, Ke R, Lo C. Doubling the spatial resolution in capacitive tactile sensors. Journal of Micro/Nanolithography, MEMS, and MOEMS. 2006;**16**(3):1-9

[23] Wu MC, Solgaard O, Ford JE. Optical MEMS for lightwave communication. Journal of Lightwave Technology. 2006; **24**(12):4433-4454

[24] Salve M, Dhone M, Rewatkar P, Balpande S, Kalambe J. Design and sensitivity analysis of micro-cantilever based biosensor for tumor detection. Sensor Letters. 2019;**17**:64-68

[25] Mehta K, Bansal D, Bajpai A, Minhas A, Kumar A, Kaur M, et al. Reduced dielectric charging RF MEMS capacitive switch. Journal of Micro/ Nanolithography, MEMS, and MOEMS. 2018;**17**(4):1-6

[26] Uppalapati PL, Malayappan B, Krishnaswamy N, Pattnaik PK. Electrostatically tunable MOEMS waveguide Bragg grating-based DWDM optical filter. Journal of Micro/ Nanolithography, MEMS, and MOEMS. 2019;**18**:015503

[27] Chu JH, Bak YH, Kang BK, Kim U, Oum KW, Choi JG, et al. Michelson interferometric detection for opto acoustic spectroscopy. Optics Communication. 1992;**89**:135-139

[28] Breguet J, Pellaux JP, Gisin N. Photoacoustic detection of trace gases with an optical microphone. Sensor and Actuators A Physics. 1995;**48**:29-35

[29] Wang W, Wu N, Tian Y, Wang X, Niezrecki C, Chen J. Optical pressure/ acoustic sensor with precise Fabry-Perot cavity length control using angle polished fiber. Optics Express. 2009;**17**: 16613-16618

[30] Ghatak A, Thyagarajan K. An Introduction to Fiber Optics. Cambridge, UK: Cambridge University Press; 1998

[31] Ranacher C et al. Characterization of evanescent field gas sensor structures based on silicon photonics. IEEE Photonics Journal. 2018;**10**(5):1-14. DOI: 10.1109/JPHOT.2018.2866628, Art no. 2700614

[32] Ranacher C, Consani C, Maier FJ, Hedenig U, Jannesari R, Lavchiev V,

et al. Spectroscopic gas sensing using a silicon slab waveguide. Procedia Engineering. 2016;**168**:1265-1269

[33] Ranacher C, Tortschanoff A, Consani C, Moridi M, Grille T, Jakoby B. Photonic gas sensor using a silicon strip waveguide. Proceedings. 2017;**1**:547. DOI: 10.3390/proceedings1040547

[34] Fan Y, Peng S, Liu Y. Transverse magnetic modes in hybrid nanoslot waveguides. Results in Physics. 2020;**16**: 103017

[35] Chandra V, Ranjan R. Performance analysis of different slot waveguide structures for evanescent field based gas sensor applications. Optical and Quantum Electronics (Springer). 2021; **53**(1–15):457. DOI: 10.1007/s11082-021-03102-8

[36] Mu J, Zhang H, Huang W. A theoretical investigation of slot waveguide Bragg gratings. IEEE Journal of Quantum Electronics. 2008;**44**: 622-627

[37] Penades JS, Khokhar AZ, Nedeljkovic M, Mashanovich GZ. Low-loss mid-infrared SOI slot waveguides. IEEE Photonics Technology Letters. 2015;**27**:1197-1199

[38] Kai-he X, Xin Z, Xiao-wei Z, Han J, Jia-wen J. A slot microring sensor with feedback spiral waveguide for trace gas CH_4 sensing in mid-infrared region. Optoelectronics Letters. 2019;**15**(1):1-5

[39] Su P, Han Z, Kita D, Becla P, Lin H, Deckoff-Jones S, et al. Monolithic on-chip mid-IR methane gas sensor with waveguide-integrated detector. Applied Physics Letter. 2019;**114**:051103

[40] Moumen S, Raible I, Krauß A, Wöllenstein J. Infrared investigation of CO_2 sorption by amine based materials for the development of a NDIR CO_2 sensor. Sensors and Actuators B: Chemical. 2016;**236**:1083-1090

[41] Fleming L, Gibson D, Song S, Li C, Reid S. Reducing N_2O induced cross-talk in a NDIR CO_2 gas sensor for breath analysis using multilayer thin film optical interference coatings. Surface and Coating Technology. 2018;**336**:9-16

[42] Barritault P, Brun M, Lartigue O, Willemin J, Ouvrier-Buffet JL, Pocas S, et al. Low power CO_2 NDIR sensing using a micro-bolometer detector and a micro-hotplate IR-source. Sensors and Actuators B: Chemical. 2013;**182**:565-570

[43] Davenport JJ, Hodgkinson J, Saffell JR, Tatam RP. Non-dispersive ultra-violet spectroscopic detection of formaldehyde gas for indoor environments. IEEE Sensors Journal. 2018;**18**:2218-2228

[44] Zheng CT, Huang JQ, Ye WL, Lv M, Dang JM, Cao TS, et al. Demonstration of a portable near-infrared CH4 detection sensor based on tunable diode laser absorption spectroscopy. Infrared Physics & Technology. 2013;**61**:306-312

[45] Dong M, Zheng C, Miao S, Zhang Y, Du Q, Wang Y, et al. Development and measurements of a mid-infrared multi-gas sensor system for CO, CO_2 and CH_4 detection. Sensors. 2017;**17**:2221

[46] Vincent TA, Gardner JW. A low cost MEMS based NDIR system for the monitoring of carbon dioxide in breath analysis at ppm levels. Sensors and Actuators B: Chemical. 2016;**236**: 954-964

Section 2

Waveguiding Media

Chapter 4

Waveguides in Magnetism and Spintronics

Rebeca Díaz Pardo and Jorge Martínez Garfias

Abstract

In this chapter, we propose to review and discuss the use of waveguides in spintronic devices and other magnetic systems as well as some emergent phenomena linked to ultrafast magnetic dynamics. Spintronic devices are expected to replace the recent nanoelectronic memories and sensors due to their efficiency in energy consumption and functionality with scalability. In the field of spintronic devices, it is required the development of magnetic thin films with a wide range of magnetic properties. To achieve the characterization of magnetic thin films, several techniques are very useful, particularly in studying magnetization dynamics described by the phenomenological Landau-Lifshitz-Gilbert equation. These techniques are developed based on key phenomena such as spin pumping, ferromagnetic resonance, the recently observed terahertz oscillations and ultrafast switching. The coupling of waveguides and microwave generators to produce radio-frequency magnetic fields and integrate them with magnetic thin films is crucial to characterize the magnetization dynamics.

Keywords: spintronics, coplanar waveguides, spin current, magnetization dynamics, magnetism

1. Introduction

This chapter shows the state of the art of the use of waveguides in different experiments in magnetism, spintronics and magnonics and gives a general view of the relevance of waveguides in the development of the areas of spintronics and magnonics. We also discuss perspectives on new technologies related to these elements.

The chapter is divided as follows: we give a brief introduction to magnetization dynamics, the phenomena of resonance and the basic characteristics of spinwaves. Afterward, we describe the type of waveguides that are commonly used in in order to excite spinwaves and generate resonance conditions in magnetic materials. Then, we describe some phenomena in spintronics where oscillating fields give rise to novel and interesting phenomena. Finally, we briefly give a non-exhaustive view of potential applications and concluding remarks.

2. Magnetic dynamics

In this section, we give a brief introduction to the principles of magnetization dynamics. We start by explaining the Landau-Lifshitz-Gilbert (LLG) equation since this expression describes the dynamics of magnetic moments generated by a magnetic field and in general, by a driving force.

We suppose that we have a ferromagnetic layer with magnetic moment per unit volume **M** and it is under an effective magnetic field $\mathbf{H}_{eff}$. The effective field has several components: the Zeeman energy H_Z, exchange interaction H_{exch}, magnetocrystalline anisotropy energy H_{anis}, and the demagnetization energy H_{demag}, among others, depending on the system [1]. All these terms sum the energy interactions of magnetic materials per unit volume. We find the lowest energy state or the equilibrium state when the magnetic moment points in the same direction as the effective field, $\mathbf{H}_{eff}$. On the other hand, if the magnetization vector **M** deviates from the equilibrium state, then a torque is exerted on the magnetic moment [2]:

$$\boldsymbol{\tau} = \mathbf{m} \times \mu_0 \mathbf{H_{eff}} \tag{1}$$

where $\mathbf{m} = |\mathbf{M}|/\mathbf{M}$ is a unitary vector in direction of **M**. Using the semiclassical approximation, we have that the relation between the magnetic moment **m** and the quantum angular momentum **L** is $\mathbf{m} = |\gamma|\mathbf{L}$ where $\gamma = g\mu_B/\hbar$ is the gyromagnetic ratio, with g is a dimensionless quantity that expresses the proportionality between the quantum angular momentum and the magnetic moment and μ_B is the Bohr magneton. Due to the conservation of angular momentum, the ratio between **L** and time t is equal to the torque: $\frac{\mathrm{d}\mathbf{L}}{\mathrm{d}t} = \boldsymbol{\tau}$. Therefore, from Eq. (1) and the definition of torque, we have [2]:

$$\frac{\mathrm{d}\mathbf{m}}{\mathrm{d}t} = -|\gamma|\mu_0(\mathbf{m} \times \mathbf{H_{eff}}) \tag{2}$$

which is the unitary version of the Landau Lifschitz (LL) equation with no damping [2]. The complete version of the dynamic description includes a damping term therefore the complete description is called the Landau-Lifschitz-Gilbert (LLG) equation:

$$\frac{\mathrm{d}\mathbf{m}}{\mathrm{d}t} = -|\gamma|\mu_0(\mathbf{m} \times \mathbf{H_{eff}}) + \alpha\left(\mathbf{m} \times \frac{\mathrm{d}\mathbf{m}}{\mathrm{d}t}\right) \tag{3}$$

The second term describes the damping in the magnetization, where the precession decays continuously. This decaying depends on the α term, this is the dissipation term that encompasses all the energy losses [3]. **Figure 1** shows the magnetization trajectory for different values of α. Without this extra damping term, **m** would precess indefinitely. The dissipation causes **m** to align parallel to $\mathbf{H}_{eff}$ its state of minimum energy.

2.1 Ferromagnetic resonance

In this section, we describe the phenomenology of ferromagnetic resonance, which is a widely used technique to characterize ferromagnetic materials. The magnetic moments of a ferromagnetic material present an intrinsic magnetic moment that

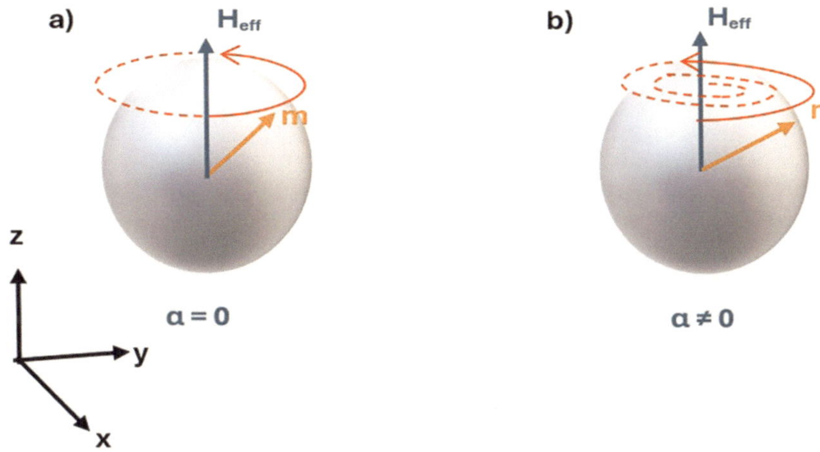

Figure 1.
Sketch of the trajectory followed by **m** *over time under* $\mathbf{H}_{eff}$ *with (a)* $\alpha = 0$ *and (b)* $\alpha \neq 0$.

precesses at a given frequency (Larmor frequency) and couples with an electromagnetic wave, absorbing its power if the frequency of the wave is the same as the Larmor frequency. This phenomenon is known as ferromagnetic resonance (FMR) [4]. This effect has various applications such as in spectroscopic techniques and in the conception of microwave devices.

Let us consider a ferromagnetic sample under a uniform external magnetic field $\mathbf{H_0}$ with an arbitrary shape. We can express the internal field $\mathbf{H_{int}}$ as [3]:

$$\mathbf{H_{int}} = \mathbf{H_0} - \mathbf{N} \cdot \mathbf{M} \quad (4)$$

with **N** the tensor of the demagnetizing factors. If we choose the Cartesian coordinate system with the x, y, and z directions along the symmetry axes of the sample as shown in **Figure 2a**, we have that the tensor is diagonal, and that its components satisfy [3]:

$$N_x + N_y + N_z = 1 \quad (5)$$

If we consider an external uniform field $\mathbf{H}_{ext}$ applied along the z axis and the magnetization precessing around it as shown in **Figure 2a**. From Eq. (4), we consider m_x and m_y, the components of the magnetization equation of motion. Solving these equations, we obtain the general Kittel resonance frequency [3]:

$$\omega_0 = \gamma\mu_0[H_0 + (N_x - N_z M)]^{1/2}\left[H_0 + \left(N_y - N_z\right)M\right]^{1/2} \quad (6)$$

The above equation shows that the FMR frequency depends on the shape of the sample and the direction of the applied field. For thin films we take the cases that are represented in **Figure 2**. For the case of the configuration of **Figure 2a**, the field $\mathbf{H}_{ext}$ is perpendicular to the sample plane, and the magnetization $\mathbf{M}_{ext}$ only acts on the z axis, so the component $N_z = 1$. Since it does not act on the x, y plane, we have that $N_x = N_y = 0$. For the configuration in **Figure 2b**, the magnetization acts on the plane of the thin film; for simplicity, we consider it acts on the direction of the x axis, that is, $N_x = 1$, and since it only acts on this axis, we have that for the axes x and y

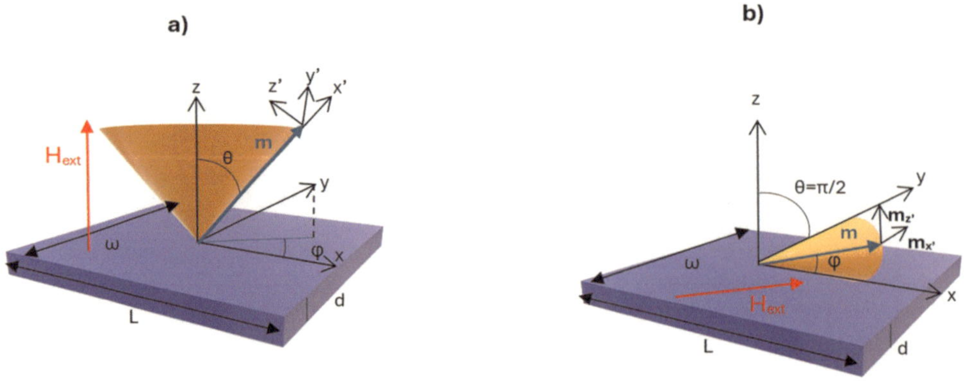

Figure 2.
*General coordinate system (x, y, z) and local coordinate system (x', y', z') where the precessing magnetization **m** is located. (a) Precession around the applied field $\mathbf{H}_{ext}$ perpendicular to the sample. (b) The transformation is performed by first rotating φ around $z = z'$ and then rotating by $(\frac{\pi}{2} - \theta)$ around the new y' axis [4]. Representation for the case when the external applied field in the sample plane, where we have the component of the magnetization m_x [5].*

$N_y = N_z = 0$. The solutions of these two configurations are represented by the following equations, respectively [3]:

$$f_{res} = (\gamma/2\pi)\left[H_0 - \mu_0 M_{eff}\right] \tag{7}$$

$$f_{res} = (\gamma/2\pi)\left[H_0(H_0 + \mu_0 M_{eff})\right]^{1/2} \tag{8}$$

With these resonance conditions, we can describe the magnetization motion by taking as reference the field with local coordinates already transformed, as in **Figure 2b**. We assume that m_x is the saturated component along the external field and therefore $m_{x'} = \frac{M_{x'}}{M_s} = 1$. If this is the case when we change the parametric description of **m** in the local coordinate system $(x', y'z')$ we perform two rotations [6]:

$$\begin{pmatrix} m_x \\ m_y \\ m_z \end{pmatrix} = \begin{pmatrix} sen\theta cos\phi & -sen\phi & -cos\theta cos\phi \\ sen\theta sen\phi & cos\phi & -cos\theta sen\phi \\ cos\theta & 0 & sen\theta \end{pmatrix} \begin{pmatrix} m_{x'} \\ m_{y'} \\ m_{z'} \end{pmatrix} \tag{9}$$

We have that in **Figure 2b**, the dynamics are limited to the $y'z'$ plane, in this case $z' = z$. If we rewrite the external field as the sum of the static field and the alternating field $H_{ext} = H + h(t)$, and replace this in the LLG equation, we can describe the dynamics of the magnetization. Moreover, we can write the magnetization components in the local coordinate system as a function of $h(t)$ through the susceptibility tensor that expresses the shape of the curve close to the resonance [5]:

$$\begin{pmatrix} m_{y'} \\ m_z \end{pmatrix} = \begin{pmatrix} \chi_{y'y'} & \chi_{y'z} \\ \chi_{zy'} & \chi_{zz} \end{pmatrix} \begin{pmatrix} h_{y'} \\ h_z \end{pmatrix} \tag{10}$$

The shape of absorbed power during the resonance is in general a combination of symmetric and antisymmetric Lorentzian functions. The components of the susceptibility tensor in Eq. (10) depend on the parameters of these Lorentzian functions [5].

The absorbed power in the resonance condition can be expressed as:

$$p_{abs}(H, H_0, \Delta H) = A(F_S(H)\cos(\varepsilon) + F_A(H)\sin(\varepsilon)) \tag{11}$$

Where A is an amplitude factor, ε is an angle that mixes the symmetric and antisymmetric Lorentzian functions F_S and $F_{A,}$ respectively [6]:

$$F_S(H) = \frac{(\Delta H)^2}{(H - H_0)^2 + (\Delta H)^2} \tag{12}$$

$$F_A(H) = \frac{\Delta H(H - H_0)}{(H - H_0)^2 + (\Delta H)^2} \tag{13}$$

The line width ΔH is half the width at half maximum in the symmetric Lorentzian curve, and the terms of the type χ denote the susceptibilities at resonance. The term ΔH is the result only of the Gilbert damping parameter α_{eff} which is composed of the interaction of H_{eff} with the resonance frequency given by the equation of Kittel as in the text [5, 6]:

$$\Delta H = \frac{\alpha_{eff}\,\omega}{\mu_0 \gamma} \tag{14}$$

Thus we have that our resonance condition FMR expresses the magnetization movement where the parameter of α_{eff} corresponds to the speed at which the spin angular momentum dissipates [6].

In order to detect the FMR signal, a microwave resonant cavity with an electromagnet is required. The resonant cavity is fixed at a frequency in the super GHz range, and a detector is placed at the end of the cavity to detect the microwaves. To achieve a full characterization of thin film materials, one needs to carry out FMR measurements at a wide range of microwave frequencies, and for this end, a cavity at a fixed frequency is not convenient. It has been shown that using a broadband microwave signal generator, a coplanar waveguide (CPW), and a broadband microwave detector, it is possible to obtain a good signal-to-noise ratio when a GHz current is injected into the CPW from one end [5, 7].

2.2 Spin waves

So far, we have considered only the case where the magnetization dynamics are homogeneous throughout all the samples. However, it is possible to find varying magnetization through the sample; the elementary excitations are known as spin waves (SW). The quanta of these excitations are known as magnons.

To describe this phenomenon, we consider a linear chain of N classical spins, each with magnitude S coupled with nearest-neighbor exchange interaction, uniformly spaced, and separated by a distance a. We consider only the Heisenberg interaction and the ground state to be where all the spins are parallel, therefore, $\mathbf{S}_i \cdot \mathbf{S}_{i+1} = \mathbf{S}^2$ and the exchange energy becomes $U_0 - 2JNS^2$ [3].

In 1930 Felix Bloch showed that the low-lying excitations of the spin system consisted of non-localized, collective spin deviations that he called SW. This was a further conceptualization from the Pierre Weiss model that considered the first excited state to be one spin reversed in the spin chain described before [8]. The existence of low-energy excitations is related to the spin system having an infinite number of degenerate ground states with infinitesimally different spin orientations in the absence of an external field.

In the linear spin chain, the effective field arising from the exchange interaction is:

$$\mathbf{H}_{exc}^{eff} = -\frac{2J}{g\mu_B\mu_0}(\mathbf{S}_{i-1} + \mathbf{S}_{i+1}) \tag{15}$$

Using the first term of the LLG equation and considering that the spins are subject to an applied static magnetic field **H**, we write the total field acting over the spins as $\mathbf{B}_T = \mu_0(\mathbf{H} + \mathbf{H}_{exc}^{eff})$. Therefore, the spin equation of motion is written as follows [3]:

$$\frac{d\mathbf{S}_i}{dt} = -\gamma\mu_0\mathbf{S}_i \times (\mathbf{H} + \mathbf{H}_{exc}^{eff}) \tag{16}$$

where $\gamma = g\mu_B/\hbar$ is the gyromagnetic ratio. Developing these equations for each component and taking into account that the magnetization points in the opposite direction of the spin and therefore taking $\mathbf{H} = -\mathbf{z}H$, we find that the possible solutions are in the form of traveling waves:

$$S_i^x = A_x e^{kx_i - \omega t}, \quad S_i^x y = A_y e^{kx_i - \omega t} \tag{17}$$

with ω the angular frequency and k the wave number. Substituting these solutions in the equations of spin motion for each component, we obtain the dispersion relation, in other words, the dependence of the spin wave frequency on the wave number:

$$\omega_k = \gamma\mu_0 H + \frac{4JS}{\hbar}(1 - \cos ka) \tag{18}$$

From here, we get the relation between the amplitudes of the spin components, and we can write the real parts of the transverse components as:

$$S_i^x = A_0 \cos(kx_i - \omega_k t), \quad S_i^y = A_0 \sin(kx_i - \omega_k t), \tag{19}$$

From these equations, we can represent the classical picture of a SW in one dimension where the spins precess circularly around the equilibrium direction, as shown in **Figure 3**. If the wave travels along the $+x$ shortest distance between two

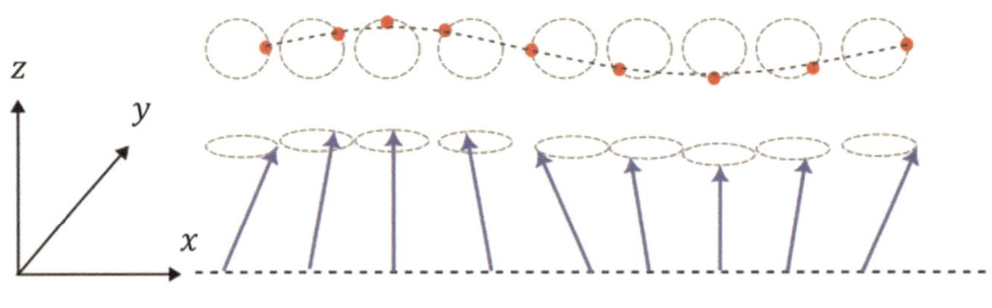

Figure 3.
Representation of a spin wave in a 1D chain propagating in the x direction. In the upper part of the figure, we show the top view of the spins. The wavelength corresponds to the distance between two spins pointing in the same direction.

spins that precess with the same phase is the wavelength corresponding the wave number $\lambda = 2\pi/k$.

When $k = 0$, the magnon energy is determined only by the magnetic field intensity $\hbar\omega_0 = \hbar\gamma\mu_0 H$, however, the general expression for the energy of one magnon in a 1D chain is:

$$\hbar\omega_k = \hbar\gamma\mu_0 H + 4JS(1 - \cos ka) \quad (20)$$

As the wave number increases, the phase difference of precession for neighboring spins. Typically, the Zeeman contributions ($\hbar\omega_0 = \hbar\gamma\mu_0 H$) of the applied magnetic fields are small in comparison with the exchange energy. For $ka = \pi$, the exchange energy is maximum and its value is $8JS$, and it is the energy of a linear chain when two consecutive spins are reversed.

At low temperatures, the modes with small wave numbers are dominant, and they are important for many magnonic phenomena. For $ka \ll 1$, we can rewrite Eq. (20) as:

$$\hbar\omega_k \approx \hbar\gamma\mu_0 H + 2JSa^2k^2 \quad (21)$$

which is a quadratic dispersion relation with an energy gap. The frequencies of SWs with $ka \ll 1$ are in the microwave range. When $k \sim 1/a$ the spin wave frequencies are in terahertz range.

There are three types of SWs based on the propagation direction (k) with respect to the external magnetic field (H_ext) and the magnetization (M): Damon-Eshbach (DE) or surface waves when **k** is perpendicular to the field **H** , but both are on the sample plane, backward volume waves (BV) when **k** is parallel to the field **H** and both are on the sample plane and forward volume waves (FV), when **k** is perpendicular to the field **H** but the field is perpendicular to the sample plane [9].

3. Coplanar waveguides in magnetic resonance experiments

Waveguides are important in the study of any high-frequency signal and since we have described the magnetization dynamics in previous sections, we continue to describe examples of applications in magnetism and spintronics. In ferromagnetic materials, the Larmor frequency usually falls in the range of a few tens of GHz. Thus, transmission lines, such as coplanar waveguides, provide a large frequency band in which the impedance is relevant in FMR-related experiments where ferromagnetic films are studied [10].

Coplanar waveguides are composed of a thin metal film layer that is the conductor with a width W_{sl} with two ground planes parallel to the conductor with a width W_{gap} separated by a distance W_{gl} which in this case are composed of copper and have a thickness t_{TL} where they are placed on a dielectric substrate with a thickness $t_{substrate}$ as shown in **Figure 4** [11].

One of the main features of coplanar waveguides is the different possible combinations of W_{gap} and W_{sl} to design circuits with a specific impedance and modulation of the electromagnetic field generated by the signal line and its two ground lines. This is of great interest to us because we use this property to control the magnetic field generated by the CPW at frequencies from 0 GHz to 20 GHz, which corresponds to the study of the properties of ferromagnetic resonance (FMR) [12].

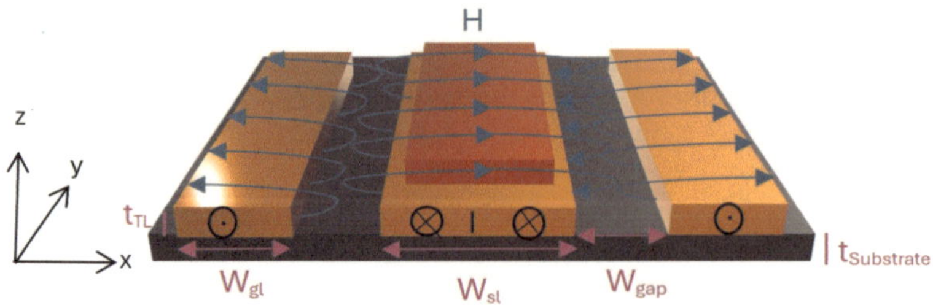

Figure 4.
Coplanar waveguide. The current j flows inside and outside along the direction y, generating the transverse magnetic field lines h(t). The most important parameters of a coplanar waveguide are indicated: the metal thickness t_{TL}, the substrate thickness $t_{substrate}$, W_{sl}, the width of the signal line, the width of the gap between the signal line and the ground lines W_{gap}, and the width of the ground line W_{gl} [10].

3.1 Theoretical model coplanar waveguide in magnetism

We give a brief theoretical description of the physics involved in the operation of CPWs. We start by assuming that electromagnetic waves propagate along the y direction of the CPW, as shown in **Figure 4**. In this CPW system, transverse electromagnetic modes (TEM) are admitted, and the local electric $\mathbf{E}^{-}$ and magnetic field $\mathbf{H}$ are perpendicular to each other. The direction of propagation and the direction of the magnetic field $\mathbf{H}$ are indicated in **Figure 4**. Applying Maxwell's equations to a general homogeneous lossy medium [10]:

$$\nabla \times \mathbf{E} = -i\omega\mu\mathbf{H} \tag{22}$$

$$\nabla \times \mathbf{H} = i\omega\varepsilon\mathbf{E} + \sigma\mathbf{E} \tag{23}$$

where $\omega/2\pi$ is the frequency of the electromagnetic wave, $\mu = \mu' + i\mu'$ is the complex magnetic permeability, $\varepsilon = \varepsilon' - i\varepsilon''$ is the complex electrical permittivity and σ is the conductivity of the guide wave [10]. For the electric field, we have to [10, 13]:

$$\nabla^2\mathbf{E} + \omega^2\mu\varepsilon\left(1 - i\frac{\sigma}{\omega\sigma}\right)\mathbf{E} = 0 \tag{24}$$

We define the complex propagation constant for the medium as follows [10]:

$$\gamma = i\omega\sqrt{\mu\varepsilon}\sqrt{1 - i\frac{\sigma}{\omega\varepsilon}} \tag{25}$$

If we assume that the electric field acts only in the x direction and, at the same time, remains uniform in the xy plane, Eq. (24) reduces to [10]:

$$\frac{\partial^2 E_x}{\partial z^2} - \gamma^2 E_x = 0 \tag{26}$$

The general solution of this differential equation is given by [10]:

$$E_x(z) = E^+\exp(-\gamma z) + E^-\exp(\gamma z) \tag{27}$$

where the term on the right side describes a wave that travels toward the positive side of the z axis and the left side of a wave that travels in the opposite direction in the

$-z$ direction. The time-dependent propagation factor of the positively moving wave (in this case $-\gamma z$) is written as [10]:

$$\Re(\exp(-\gamma z)\exp(i\omega t)) = \Re(\exp(-\alpha z)\exp(-i\beta z)\exp(i\omega t)) = \exp(-\alpha z)\cos(\omega t - \beta z) \tag{28}$$

For good conductors $\omega < < \sigma$, this means that the dielectric losses are small compared to the conductivity, and when the conductivity loss predominates, we have that Eq. (25) reduces to [10]:

$$\gamma = i\omega\sqrt{\mu\varepsilon}\sqrt{\frac{\sigma}{j\omega\varepsilon}} = (1+i)\sqrt{\frac{\omega\mu\sigma}{2}} \tag{29}$$

Where the real part of the above equation corresponds to α and this defines the skin depth δ_s of the electric field in the material with the relationship [10]:

$$\delta_s = \frac{1}{\alpha} = \sqrt{\frac{2}{\omega\mu\sigma}} \tag{30}$$

where $\alpha = i\omega\sqrt{\mu\varepsilon}$. As we can see from the above equation, the amplitude of the current density decreases exponentially on the scale of the characteristic length δ_s within conductive transmission lines [10].

This result is related to the geometric parameters of CPW. If $t_{Tl} < < \delta_s$ we approximate the current density to be homogeneously distributed along the z axis in the CPW. In contrast, along the x direction and due to W_{sl}, $W_{gl} > > \delta_s$ we cannot assume that the current density is homogeneous. Thus, to determine the local in-plane and out-of-plane amplitudes of the excitation fields, one has to consider the distribution of the current density layers as a function of frequency along the x direction [10, 13].

Thus, we have that the change in the increase in current densities that goes toward the limits of the transmission lines is proportional to the frequency, where they are related through the depth of the skin δ. Therefore, the frequency-dependent magnetic field within the CPW can now be calculated by summing all contributions from the current density paths using the Biot-Savart law. Thus, the field can be calculated at each point of the CPW [10].

3.2 Example of CPW compatible with FMR experiment

As explained in the previous section, a coplanar waveguide (CPW), as shown in **Figure 4**, consists of ground lines and conducting lines. The ground lines must be at ground potential, and the signal line must have relative voltage amplitude [10, 13]. We notice in **Figure 4** that the magnetic field lines are parallel to axis x and if we put a thin ferromagnetic film on top, these field lines would lie on the sample plane [10].

Now, we will discuss an example of a CPW design, starting with the choice of materials for an FMR experiment. The conductive material that was used is copper. This metal is commonly chosen due to its low cost, good electrical conductivity and low dielectric loss. It is also compatible with the transmission of the wave frequencies studied in an adequate way. The substrate we use is Rogers RO4350B, which is made of glass-reinforced woven hydrocarbon/ceramic with a dielectric constant of $\varepsilon_r = 3.48 \pm 0.05$ [14]. According to the data we have, this material has a high frequency for applications of up to 18–22 GHz with a loss of ± 0.05. Moreover, its dielectric constant does not change abruptly with temperature therefore is a suitable material for temperature variation experiments [14].

To design the waveguide, we use the Advancing the Wireless Revolution Design Environment (AWR) program, which is provided under license by Cadence Design System, Inc. [15].

The process to use this software is to delimit the parameters. For this purpose, we use the AWR transmission line calculator (TX-LINE) to get a first approximation of realistic parameters (considering an impedance of 50Ω and a loss close to 0) [15]. Then, in circuit schematics, using the predefined designs and the dimensions of the CPW, we create an electromagnetic (EM) structure. At this stage, we design the CPW in three-dimensions (3D) and analyze how the electromagnetic fields act on the waveguide and how they affect the loss and power in it. What this procedure is broadly about is comparing the design in **Figure 5b** with the ideal dimensions that we consider in **Figure 5a**, finding the best configuration, following this procedure.

Furthermore, the CPW is designed in 3D, taking into account the conductor and substrate mentioned above in the Axiem EM section of AWR [15]. This software simulates the electromagnetic fields in the CPW in order to calculate the loss and power of our reference CPW, which is shown in **Figure 5b** and compared to the loss and power in the CPW with the parameters obtained by TX-LINE. Thus, by the aforementioned procedure, we refine our design until we reach the one we have graphed in **Figure 5a**, which minimizes the loss and maximizes the power (**Figure 5a**).

We see how the loss of both the approximate CPW and that of the CPW designed in 3D is very close to zero, and the power of the CPW designed in 3D (EM Structure) is delimited below the CPW that we take as a base (Coplanar), this tells us that we have a suitable design for our purposes. Thus, after this process, we have the dimensions found for this example of CPW using the characteristics of **Figure 4** are: $W_{gl} = 3mm$, $W_{sl} = 1mm$, $W_{gap} = 0.2mm$, $t_{TL} = 0.2mm$.

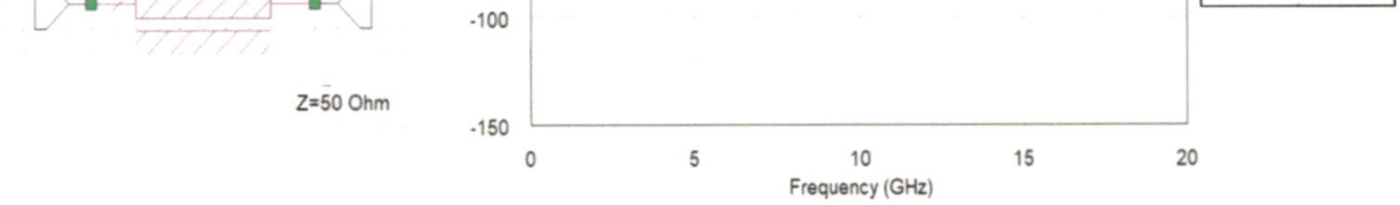

Figure 5.
The coplanar waveguide layout of our example designed in AWR is shown [15]. (a) Circuit diagram for the designed coplanar waveguide. (b) Return loss and power vs. frequency. The pink squares represent the loss of the CPW modeled in the first approximation with the TXline parameters, the brown diamonds represent the loss of the AWR-designed CPW, the blue triangles represent the power of the approximate CPW, and the red double triangles represent the power of the AWR-designed CPW.

4. Spintronics and magnonics phenomena with waveguides

In this section, we describe experiments and physical phenomena that are intensely studied in the subareas of spintronics and magnetization dynamics. We will focus on potential applications based on the production of spin currents, phenomena related to spin-orbit torques and spin transport via spinwaves.

4.1 FMR, spin pumping and spin currents

The magnetization dynamics in thin magnetic multilayers, specifically in ferromagnetic films in contact with paramagnetic conductors, can produce a moving magnetization vector that causes "pumping" of spins into adjacent nonmagnetic layers [16]. This spin transfer affects the magnetization dynamics, similar to the Landau-Lifshitz-Gilbert phenomenology. This phenomenon is known as spin pumping, and it is the Onsager reciprocal to spin-transfer torque (STT) [17]. In both cases, there is transfer of spin angular momentum between a ferromagnet (F) and a metallic reservoir.

In the first one, a spin (polarized) current injected from the metal into the ferromagnet transfers angular momentum onto the magnetization M of the latter, thereby inducing magnetization dynamics [18]. In contrast, for spin pumping, the nonequilibrium magnetization dynamics (magnons) in the ferromagnet act as the source for an angular momentum flow from the ferromagnet into the metal, inducing a spin current [19].

There are two different detection schemes for spin pumping in ferromagnetic/metallic (F/M) bilayers: the first one detects spin pumping spectroscopically by the increase of the (Gilbert) damping constant of coherent magnetization precession in the ferromagnet, in this case, the metal acts like an angular momentum sink. The second is the electrical detection of the spin current electrically via the inverse spin Hall effect in the metal, which acts as a spin current detector. The excedent of angular momentum can also flow out or be "pumped" across the interface from the ferromagnet into the normal metal, where both longitudinal and transversal angular momentum are pumped. The latter is the main contributor to damping for small precession angles θ. This is schematically shown in **Figure 6d**. Since spin pumping increases the damping α, one way to detect spin pumping is through the increase in the line width in FMR measurements, as compared in **Figure 6a** and **b**.

The spin current density (the spin current per interface area) pumped across the F/N interface by the spin-pumping mechanism is [20–22]:

$$\mathbf{j}_s^{pump} = \frac{\hbar}{4\pi}\left\{ Re\left(g^{\uparrow\downarrow}\right)\left[\mathbf{m} \times \frac{d\mathbf{m}}{dt}\right] - Im\left(g^{\uparrow\downarrow}\right)\frac{d\mathbf{m}}{dt}\right\}. \quad (31)$$

where $g^{\uparrow\downarrow}$ is the spin mixing conductance that can be understood as the available spin transport channels at the interface, $\mathbf{m} = \mathbf{M}/|\mathbf{M}|$ is the magnetization unit vector and $\hbar$ is the reduced Planck constant.

We can see from Eq. (31) that the current generated as a result of spin pumping across the F/N interface is equivalent to an additional Gilbert-like damping channel, increasing the FMR line width. The additional Gilbert-like damping arising from spin pumping can thus be expressed as an increase of α by α_{SP} [23, 24]:

$$\alpha_{SP} = g\mu_B \frac{Re\left(g^{\uparrow\downarrow}\right)}{4\pi M_S}\frac{1}{t_F}\eta \quad (32)$$

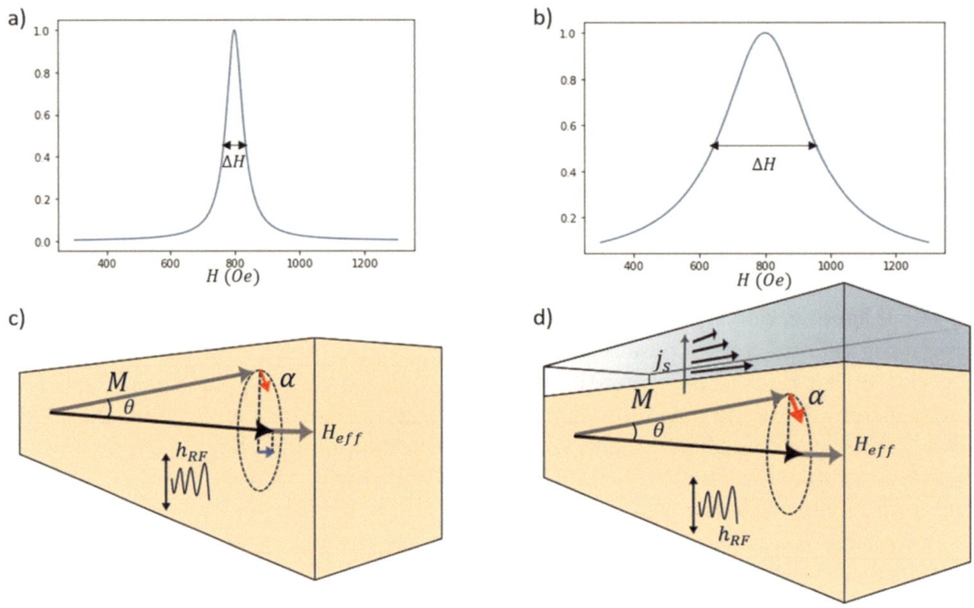

Figure 6.
(a) The absorption in the resonance condition corresponding to (b). The linewidth ΔH is determined by all the relaxation mechanisms, (c) the magnetization vector of a ferromagnet out of its equilibrium position precessing around the magnetic field with an angle θ and excited via resonance produced by a microwave magnetic field h_{RF}, and (d) ferromagnet/normal metal bilayer. In this structure, there is an additional component to the damping mechanism due to the excess of angular momentum that broads the FMR line width. This can be understood as the generation of pure spin current j_s with longitudinal spin-polarization; that is, the spin current is pumped from the ferromagnet in its resonance condition into the normal metal.

where g Landé g-value, the Bohr magneton μ_B, the saturation magnetization of the ferromagnet M_S, the thickness of the ferromagnet t_F, and the backflow parameter $0 \leq \eta \leq 1$, account for a possible spin current backflow into the ferromagnet.

Another way to detect spin currents arising from spin pumping is via electrical measurements through the inverse spin Hall effect (ISHE) [25, 26]. The charge current density can be written as:

$$\mathbf{j}_c = -\frac{2e}{\hbar}\Theta_{SH}\mathbf{j}_s \times \mathbf{s} \tag{33}$$

where $\mathbf{j}_s$ is the spin current in conductors with nonzero spin-orbit coupling and e is the electronic charge, $\mathbf{s}$ is the spin current polarization and represents the orientation of the spin flow in the direction of $\mathbf{j}_S$ and Θ_{SH} is the spin Hall angle, which represents the spin-to-charge current conversion efficiency; for heavy metals, we expect large spin-to-charge conversion.

From Eq. (31), we can see that the spin current $\mathbf{j}_S$ has AC and DC components. The DC spin current component arises from the longitudinal nonequilibrium magnetization components and the AC spin current components from the transverse components.

From Eq. (31), we can see that the spin current $\mathbf{j}_S$ has AC and DC components. The DC spin current component arises from the longitudinal nonequilibrium magnetization components and the AC spin current components from the transverse components.

For the electrically detected spin pumping, we will consider only the DC part of $\mathbf{j}_s$. The DC $\mathbf{j}_s$ is pumped across the surface and, depending on the direction of **M** and according to Eq. (33), $\mathbf{j}_c$ lies along the N plane, as shown in **Figure 7**. The pumped DC spin current $\mathbf{j}_s$ propagates across the F/N interface and is normal to it; this is the physical origin of the ISHE. From Eq. (33), we see that $\mathbf{j}_c$ lies in the plane of the N layer and perpendicular to **s**. The pumped spin current into N decays at the same scale as the spin diffusion length λ_{sp}, usually between 1 and 100 nm.

It is important to notice that $\mathbf{j}_s$ will be pumped into N within the resonance condition. Under these conditions, it is possible to detect $j_c = |\mathbf{j}_c|$ or $E_c = V_{DC}/L = j_c/\sigma_{tot}$ where L is the distance between the electrical contacts and σ_{tot} is the total conductivity of the F/N bilayer. The spin-pumping spin Hall voltage V_{DC} is given by [27, 28]:

$$\frac{V_{DC}}{L} = \frac{2e}{\hbar}\Theta_{SH}\frac{j_s\eta\lambda_{sd}\,tanh\left(\frac{t_N}{2\lambda_{sd}}\right)}{\sigma_F t_F + \sigma_N t_N} \tag{34}$$

The DC electric field can also be generated by microwave rectification phenomena apart from ISHE. In this context, the resonant magnetization precession in FMR generates a modulation of the sample's resistivity tensor at the microwave frequency. The resistivity modulations rectify microwave-frequency charge currents induced in the sample by the microwave signal. Since different components of the resistivity tensor (anisotropic magnetoresistance, anomalous Hall effect, etc.) can contribute to the rectification voltages, based on symmetry or line shape can be ambiguous the presence of an additional, magnetically ordered layer may provide an additional magnetization damping channel [29]. Furthermore, there are other effects that should be taken into account in microwave rectification or magneto-caloric effects such as the anomalous Nernst effect, which may then play a role in the proximity of polarized metal.

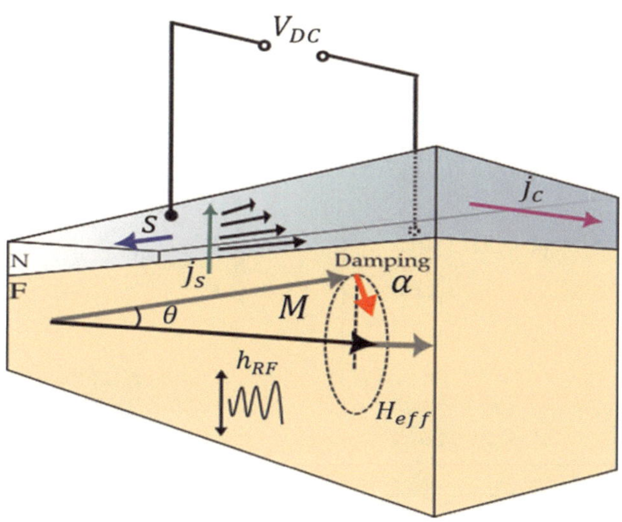

Figure 7.
Schematic of the DC electrically detected spin-pumping experiments, the DC spin current $\mathbf{j}_s$ pumped into the normal metal is converted into a charge current $\mathbf{j}_c$ via the inverse spin Hall effect and then detected via the corresponding open-circuit voltage V_{DC}.

Note that the ferromagnetic layer in this type of experiment does not need to be metallic in order to produce a spin current in the adjacent layer. There has been experimental evidence both for and against static magnetic polarization in the Pt in F/Pt hybrids made from an electrically insulating ferromagnet such as yttrium iron garnet (YIG) [30–32].

Spin pumping, as well as other interesting phenomena well known in the spintronics areas such as spin Hall effect (SHE) [25, 33], Rashba effect [34, 35], and Dyaloshinskii-Moriya Interaction (DMI) are all known to have origin in spin-orbit coupling (SOC) [36].

Spin-orbit torques (SOT) originating from pure spin current are expected to be crucial to control the magnetization as well as SW much more efficiently than for instance conventional spin-transfer torque, since SOT spin-polarization arises from the carrier velocity difference, and there is no need for a magnetized layer that plays as a spin filter. SOT is capable of compensating Gilbert damping over extended regions, which has opened up a new avenue for on-chip SW communication devices. These applications will be discussed further in Section 5.

We have discussed spin-pumping experiments in which the magnetization was driven out of equilibrium via conventional FMR. In other words, the spin current across the F/N interface was excited by microwave photons. However, magnetization dynamics can be caused by other mechanisms: thermal gradients, acoustic strain fields, charge currents, and high-frequency elastic deformations.

4.2 Microwave techniques to study spin waves

In magnonic device technology, the main mechanisms under study are the ones implicated in three major blocks: generation, detection, and manipulation of the spin waves, which correspond to input, gate, and output. The magnetic waveguide in a magnonic device guides the spin wave signal of a given frequency , wave vector, phase, and amplitude, which are the key ingredients for wave-based computing [37].

An important breakthrough was the understanding of the charge current to spin current conversion and vice versa. The charge-to-spin current has been achieved via SHE, where a charge current is injected into a heavy metal with a large SOC, leading to an orthogonal spin current, as explained in the previous section. The inverse mechanism ISHE is useful to detect spin waves (**Figure** 7) [33, 38].

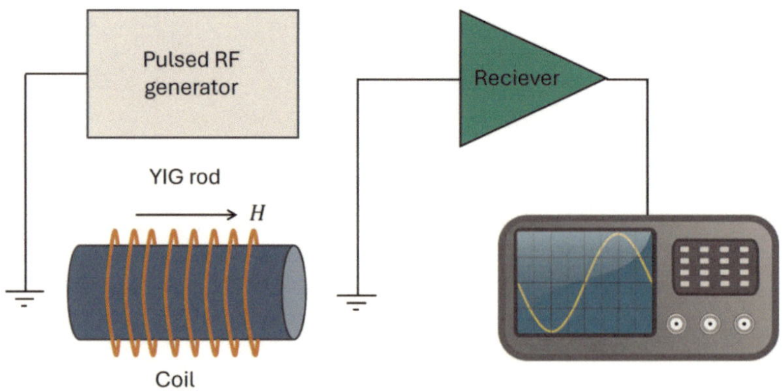

Figure 8.
Diagram of the setup used to observe spin wave frequency shift and magnon-phonon conversion in a magnetic field varying with time.

In order to propagate a magnon current, we need to be able to regulate the input, output, and gate. To do so, it is necessary to manipulate and control regimes of frequency, wavevector, and phase of the spin wave. We also need to be able to control the decay length (λ), which, as explained before, is the length over which the spin wave intensity drops by a factor of e. For this, we need materials with low damping as YIG, as we already have pointed out; other examples include Permalloy ($Ni_{80}Fe_{20}$) [39], CoFeB [40], and Heusler Alloys [41].

A very common technique to study long spin waves in ferromagnets is the use of microwave radiation. For a monochromatic driving field with frequency ω, the absorbed microwave power by a spin wave when $\omega_k = \omega$ is given by [3]:

$$P = \frac{1}{2}\mu_0\omega\Re\left[\int \mathbf{h}\cdot\mathbf{m}^* d\mathbf{r}\right] \tag{35}$$

where $\mathbf{h}$ and $\mathbf{m}$ are the driving field and the small signal magnetization with frequency ω, they are both complex quantities. If we consider the driving field along the x direction, we can write the power absorbed by a spin wave with magnetization $m_x(\mathbf{k})\exp(i\mathbf{k}\cdot\mathbf{r})$ due to a driving field with a spatial variation $h_x \exp(i\mathbf{q}\cdot\mathbf{r})$ as:

$$P = \frac{1}{2}\mu_0\omega V\Re\left[h_x(\mathbf{q})m_x^*(\mathbf{k})\right]\delta_{\mathbf{q},\mathbf{k}} = \frac{1}{2}\mu_0\omega V\chi''_{xx}|h_x(\mathbf{k})|^2 \tag{36}$$

where $\delta_{\mathbf{q},\mathbf{k}}$ is the Kronecker delta, equal to 1 for $\mathbf{q} = \mathbf{k}$ and zero for other values, χ''_{xx} is the susceptibility tensor. This expression shows that a spin wave with a certain wave vector $\mathbf{k}$ can only be excited by a driving magnetic field with a Fourier component at the same wave vector.

In bulk materials with uniform internal magnetizing fields, it is only possible to excite the $k \sim 0$ magnon, that is, the case of ferromagnetic resonance.

In order to excite higher-order magnons, we need microwave radiation, fine wire antennas, and thin magnetic films. For instance, a metallic multielement antenna is deposited by lithography on the surface of the magnetic film [42].

Another possibility is to use the boundary conditions at the film surface in order to generate a nonuniform internal driving field. In this scheme, the driven spin waves are comparable to the film thickness [43]. In films, the first observations of spin wave resonances were measured in a Permalloy film, with the static field perpendicular to the film surface. In this experiment, plane spin waves with wave vectors perpendicular to the film plane generated standing waves between the two film surfaces [43].

The wavelength is given by $d = n(\lambda/2)$ where d is the film thickness and n is an integer. As the field increases, the wave number varies, and the spin wave resonances produce absorption peaks corresponding to a different wavelength, with index number n.

The first studies of spin waves were done in YIG bulk single crystals in the shape of disks, rods, or slabs. When the sample has these shapes, the applied uniform field produces a nonuniform internal field due to demagnetizing effects on the surfaces. For static fields, the wave frequency remains unchanged; however, the wave number varies since there is no momentum conservation. The experiment described in ref. [44], uses a YIG displaced in a cylindrical microwave cavity with a 9.4 GHz microwave input. In the experiment performed by Eshbach et al., the oscilloscope signal showed magneto-elastic wave echoes at various magnetic field values. This indicated that due to the nonuniform *dc* field, the microwave pulse drives the spin wave in the region where it has a wave number $k \sim 0$.

The wave packet with circular wavefront propagates toward the center of the disk with constant frequency. As the field decreases, the wave number increases and approaches a crossover region. In a uniform field, the two branches of magnetoelastic dispersion would correspond to normal modes. Since the field varies in space, the excitation can hop into another branch [45]. The hopping probability depends on the combination of magnetic field values at a given position and if, in that position, the magnon and phonon wave numbers are equal. On the other hand, if the field gradient is much smaller, the probability for branch hopping decreases, and in this case, the spin wave packet is converted very efficiently into an elastic wave, just like in the experiment performed in Ref. [44].

Magnon–phonon conversion in a time-varying field can be experimentally demonstrated in a YIG rod using a setup consisting of the microwave source, that produces *rf* bursts with frequency in the range of 1–2 GHz and duration $\sim 0.2\mu s$, and the receiving system, which consists of a broadband microwave amplifier, a diode detector or a narrowband superheterodyne system, composed of a local oscillator, mixer, amplifier and diode detector.

In this type of experiment, one can tune the amplifier and measure the pulse frequency by scanning the oscillator frequency in the narrowband. Then, an oscilloscope displays the pulses and the waveform of the current pulses produced by the time-varying field [46].

We have discussed the first experiments in order to measure spin wave frequency shift and magnon–phonon conversion under a time-varying magnetic field; however, there are other measuring techniques that have recently become very important in the study of spin waves, such as inelastic light scattering techniques that nowadays are crucial to study elementary excitations in solids, among which one of the most used ones is Brillouin light scattering (**Figure 8**) [3].

5. Potential applications

Today's computing systems rely on information read and written by electric charge or voltage, and therefore, computation is performed by charge movements. The fundamental circuit element is the transistor, which plays the role of both a switch and an amplifier. Nowadays, large-scale integrated circuits are based on complementary metal–oxide–semiconductor (CMOS) field-effect transistors due to their high density, low power consumption, and low fabrication cost [47].

Logic gates can be built with CMOS transistors, such that they represent a full set of Boolean algebraic operations. The device density and the performance of the CMOS have been steadily improved in the last decades according to Moore's law [48]. CMOS technology has driven and adapted to an exploding information technology market and therefore CMOS has consolidated as a preferred mechanism in the digital domain.

Spintronics, which uses the spin degree of freedom for information coding, is particularly promising as an alternative CMOS paradigm in computing technology due to the low intrinsic energies of magnetic excitations as well as their collective nature [49–51].

Several implementations of spintronic Boolean logic devices have been investigated such as magnetic semiconductors [52], spin currents [53], nanomagnets [54], domain walls [55, 56], and skyrmions [57]. Beyond logic gates, interconnects are the key elements of any circuit. Usually more energy is typically dissipated in interconnects by moving data around. The fraction of the power dissipated in interconnects may be even larger for spin-based logic if communication happens via spin currents, domain wall propagation, or by converting back-and-forth spin signals to electrical signals. In spin waves, there is no charge transported; therefore, using them as information carriers is a potential solution for energy dissipation [58].

For this reason, of the most novel spintronic logic device concepts have been based on spin waves as information carriers, and several spin-wave-based logic devices have been proposed [47] and experimentally demonstrated, for instance, via interferometer-based logic gates [59] and spin-wave majority gates [60]. However, in order to develop practical applications of such logic devices, a combination of circuits and computing systems containing both logic and memory is necessary.

The current challenge in magnonic circuits is to keep losses under control. If this is achieved, communication may not require extra energy costs. Magnonic circuits can scale to nanoscale dimensions, unlike optical waveguides. For now, some realistic proposals point toward a hybrid systems concept with local spin-wave islands embedded in a CMOS periphery.

Another advantage of spin waves is their relatively large coherence length l_{pl}, even at room temperature (up to 1 cm in YIG of micrometer thicknesses) as well as their broad frequency range. These characteristics make spin wave devices promising candidates for applications related to data transfer and processing for *rf* applications. Additionally, the recent miniaturization of spin-wave wavelength λ below 100 nm that, for now, is limited to the lattice constant of the material gives further perspective to enable access to magnonics in the THz range [61].

Current technology covers the frequency range of up to 100 GHz, which is of special interest for the upcoming 5G communication systems. As a result, typical components for wireless communication (antennas, amplifiers, filters, mixers, modulators, and frequency synthesizers) will have to adapt to new frequency bands and communication protocols.

The development of ultrahigh-speed beyond–CMOS spin-wave-based technology for signal processing requires, as said before, long-range enhancement of spin waves, that is, enhanced spin wave coherence and propagation length. These can be addressed by spin-orbit torques (SOT) since, through these torques, it is possible to electrically control damping, resulting in significantly enhanced spin-wave propagation.

The theoretical concepts of SOC and spinwaves were both formulated separately. However, the interplay between these two topics has drawn attention recently due to its various advantages [61, 62]. There are several phenomena that originate from SOT, for instance, perpendicular magnetic anisotropy (PMA), SHE, Rashba effect, spin pumping and Dyaloshinskii-Moriya Interaction (DMI) [63, 64]. At the same time, it has recently and experimentally realized that SOT originates from pure spin current, as it is explained in Section 4.1.

Observation of chiral spin textures is caused, among other factors, by the presence of DMI in thin film heterostructures where a ferromagnetic layer is placed adjacent to a heavy metal layer or 2D material. The influence of DMI or antisymmetrical interaction in the spin wave dispersion has become a topic of interest. It has been theoretically predicted that spin wave asymmetry can be induced by DMI, and this can be measured by the frequency difference of counter-propagating spin waves in the Brillouin light scattering technique [62, 65].

SOTs provide a versatile tool to electrically manipulate the magnetization of all classes of magnetic materials, that is, metals, semiconductors, or insulators, as well as different types of magnetic order, including ferrimagnetic and antiferromagnetic structures. This versatility has led to a variety of experimental and conceptual results, which have greatly expanded the scope of spintronics in the last decade [50].

The demonstration of the current-induced switching of a single-layer ferromagnet, by injecting a current density $\sim 108A/cm^2$ into a Pt layer a few nm thick, adjacent to a Co dot with perpendicular magnetization [36]. Further experiments showed switching of three terminal magnetic tunnel junctions (MTJs) based on Ta/CoFeB/MgO or W/CoFeB/MgO stacks with either perpendicular or in-plane magnetization [66, 67], using *ns* current pulses with extremely low error rates and without external fields. In MTJs, SOTs allow for the switching of the free layer without passing a current through the tunnel barrier. The separation of the write and read current paths further avoids write errors during readout and voltage breakdowns. For these reasons, SOTs are attracting increasing attention as a replacement for or addition to STT in magnetic random access memories (MRAMs).

The goal of the scientific challenges is to combine a number of spin-orbit effects with the magnonic crystals and therefore to the magnonic devices. For example, the simultaneous existence of PMA and DMI as well as patterning of 2D magnonic crystals may stabilize topological magnetic textures like skyrmions and might improve the control of their dynamics.

A new path for novel physical phenomena may be paved by the combined effect of current flow and magnon on skyrmions due to the competition between thermally driven magnon current, transverse magnon current, and electron flow. One of the main challenges of magnonics for applications is the large damping of spinwaves in metallic ferromagnets, which limits its propagation distance and efficiency. The properties of SOT may indeed dramatically reduce damping.

6. Conclusions

Waveguides in spintronic phenomena are crucial due to the nature of magnetization dynamics, which makes magnetic moments sensitive to time-varying fields, producing a wide variety of spin phenomena from spin waves to resonance phenomena which, with proper adjacent layers, produce pure spin currents due to SHE or Rashba effect. The optimization of waveguides so that the magnetic moments reach the resonant condition is a key ingredient in generating spin currents and observing phenomena like spin pumping, SHE, and ISHE, which arise from SOC and produce SOTs.

SOTs provide a variety of tools to electrically manipulate the magnetization of all classes of magnetic materials: metals, semiconductors, or insulators, as well as different types of magnetic order (ferrimagnetic and antiferromagnetic structures). This has produced an enormous amount of experimental results, which have greatly expanded the scope of spintronics in the last decade.

Spin waves are considered to be promising as information carriers in low-power-consuming computing and logic. Since, in the case of spin waves, there is no charge transport but purely spin transport, and it presents interesting advantages such as the absence of Joule heating and clock frequencies up to the THz frequency regime.

On the other hand, the wide variety of physical phenomena and properties of materials linked to SOTs open the door for innovative technology that can leverage the topological properties of certain materials in spintronic devices, spin logic, efficient reading and writing mechanisms, etc.

One of the current challenges in the field of spintronics-magnonics lies in implementing spin-orbit effects to modify the magnon properties in tailored magnetic materials. This achievement is very promising for energy-efficient high-frequency nanoelectronic devices.

Conflict of interest

The authors declare no conflict of interest.

Abbreviations

LLG	Landau-Lifshitz-Gilbert
FMR	ferromagnetic resonance
CPW	coplanar waveguide
DE	Damon-Eshbach
TEM	transverse electromagnetic
EM	electromagnetic
STT	spin-transfer torque
ISHE	inverse spin Hall effect
YIG	ytrium iron garnet
SHE	spin Hall effect
DMI	Dyaloshinskii-Moriya interaction
SOC	spin-orbit coupling
SOT	spin-orbit torque
SW	spin wave
CMOS	complementary metal-oxide-semiconductor
PMA	perpendicular magnetic anisotropy
MTJ	magnetic tunnel junction
MRAM	magnetic random access memory

Author details

Rebeca Díaz Pardo*† and Jorge Martínez Garfias†
Instituto de Física, Universidad Nacional Autónoma de México, Ciudad de México, Mexico

*Address all correspondence to: diaz_pardo@fisica.unam.mx

† These authors contributed equally.

References

[1] Coey J. Magnetism and Magnetic Materials. The Edinburgh Building, Cambridge CB2, UK: Cambridge University Press; 2009. DOI: 10.1017/CBO9780511845000

[2] Xu J, Jin L, Liao Z, Wang Q, Tang X, Zhong Z, et al. Quantum spin-wave materials, interface effects and functional devices for information applications. Frontiers in Materials. 2020;**7**:2-6. Available from: https://www.frontiersin.org/articles/10.3389/fmats.2020.594386

[3] Rezende SM. Fundamentals of Magnonics. Cham: Springer; 2020. DOI: 10.1088/1361-6463/aac7b5

[4] Moura A. Application of magnon coherent states in spintronics. Journal of Magnetism and Magnetic Materials. 2020;**498**:166091. Available from: https://www.sciencedirect.com/science/article/pii/S0304885319329920

[5] Decker M. Spin current induced control of magnetization dynamics [Ph.D. dissertation]. Universität Regensburg; 2017. DOI: 10.5283/epub.33381. Available from: https://epub.uni-regensburg.de/37166/1/PHD_Decker_Martin_2017.pdf

[6] Pietanesi L. Spin-to-charge conversion in $Bi_2Se_3/NiFe$ heterostructures [master thesis]. Politecnico di Milano; 2020. Available from: https://www.politesi.polimi.it/handle/10589/151301

[7] Montoya E, McKinnon T, Zamani A, Girt E, Heinrich B. Broadband ferromagnetic resonance system and methods for ultrathin magnetic films. Journal of Magnetism and Magnetic Materials. 2014;**356**:12-20. Available from: https://www.sciencedirect.com/science/article/pii/S0304885313009256

[8] Bloch F. Zur theorie des ferromagnetismus. Zeitschrift für Physik. 1930;**61**:206-219

[9] Prabhakar A, Stancil DD. Spin Waves: Theory and Applications. Springer; 2010

[10] Obstbaum M. Inverse spin Hall effect in metallic heterostructures [Ph.D dissertation]. Regensburg Universität; 2015. Available from: https://epub.uni-regensburg.de/33381/1/Doktorarbeit%20Martin%20Obstbaum.pdf

[11] Simons RN. Coplanar Wavegude Circuits, Components, and Systems. John Wiley and Sons, Inc.; 2001. ISBN: 978-0-471-46393-1

[12] Pizarro DCP. Estudio de la respuesta en frecuencia de líneas de guía de onda coplanar en función de diferentes tipos de estructuras periódicas ranuradas en el plano de tierra. Centro de Investigación CIentífica y de Educación Superior De Ensenada (CICESE); 2008. Available from: https://cicese.repositorioinstitucional.mx/jspui/bitstream/1007/184/1/179031.pdf

[13] Pozar DM. Microwave Engineering. John Wiley and Sons Inc.; 2012. ISBN: 978-0-470-63155-3

[14] *RO*4359*B* Laminates. Rogers Corporation; 2024. Available from: https://www.rogerscorp.com/advanced-electronics-solutions/ro4000-series-laminates/ro4350b-laminates

[15] AWR Design Environment. Cadence Design Systems, Inc., version 17, 2024. Available from: https://www.cadence.com/en_US/home/tools/

system-analysis/rf-microwave-design.html

[16] Althammer MS, Weiler M. Spin Pumping. John Wiley and Sons, Ltd; 2015;**6**:111-124. DOI: 10.5772/intechopen.115052. Available from: https://onlinelibrary.wiley.com/doi/abs/10.1002/9781118751886.ch6

[17] Ralph D, Stiles M. Spin transfer torques. Journal of Magnetism and Magnetic Materials. 2008;**320**(7):1190-1216. Available from: https://www.sciencedirect.com/science/article/pii/S0304885307010116

[18] Liu L, Moriyama T, Ralph DC, Buhrman RA. Spin-torque ferromagnetic resonance induced by the spin hall effect. Physical Review Letters. 2011; **106**:036601. DOI: 10.1103/PhysRevLett.106.036601

[19] Wang Y, Ramaswamy R, Yang H. FMR-related phenomena in spintronic devices. Journal of Physics D: Applied Physics. 2018;**51**:273. DOI: 10.1088/1361-6463/aac7b5

[20] Tserkovnyak Y, Brataas A, Bauer GEW. Enhanced gilbert damping in thin ferromagnetic films. Physical Review Letters. 2002;**88**:117601. DOI: 10.1103/PhysRevLett.88.117601

[21] Tserkovnyak Y, Brataas A, Bauer WB, Gerrit E. Spin pumping and magnetization dynamics in metallic multilayers. Physical Review B. 2002;**66**: 224403. DOI: 10.1103/PhysRevB.66.224403

[22] Tserkovnyak Y, Brataas A, Bauer GEW, Halperin BI. Nonlocal magnetization dynamics in ferromagnetic heterostructures. Reviews of Modern Physics. 2005;**77**:1375-1421. DOI: 10.1103/RevModPhys.77.1375

[23] Jiao H, Bauer GEW. Spin backflow and ac voltage generation by spin pumping and the inverse spin hall effect. Physical Review Letters. 2013;**110**: 217602. DOI: 10.1103/PhysRevLett.110.217602

[24] Chen YT, Takahashi S, Nakayama H, Althammer M, Goennenwein STB, Saitoh E, et al. Theory of spin hall magnetoresistance. Physical Review B. 2013;**87**:144411. DOI: 10.1103/PhysRevB.87.144411

[25] Dyakonov M, Perel V. Current-induced spin orientation of electrons in semiconductors. Physics Letters A. 1971; **35**(6):459-460. Available from: https://www.sciencedirect.com/science/article/pii/0375960171901964

[26] Valenzuela S, Tinkham M. Direct electronic measurement of the spin hall effect. Nature. 2006;**442**:176-179. Available from: https://www.nature.com/articles/nature04937

[27] Mosendz O, Pearson JE, Fradin FY, Bauer GEW, Bader SD, Hoffmann A. Quantifying spin hall angles from spin pumping: Experiments and theory. Physical Review Letters. 2010;**104**: 046601. DOI: 10.1103/PhysRevLett.104.046601

[28] Mosendz O, Vlaminck V, Pearson JE, Fradin FY, Bauer GEW, Bader SD, et al. Detection and quantification of inverse spin hall effect from spin pumping in permalloy/normal metal bilayers. Physical Review B. 2010; **82**:214403. DOI: 10.1103/PhysRevB.82.214403

[29] Sun KY, Chang H, Kabatek M, Song YY, Wang Z, Jantz M, et al. Damping in yttrium iron garnet nanoscale films capped by platinum. Physical Review Letters. 2013;**111**:

106601. DOI: 10.1103/PhysRevLett.111.106601

[30] Geprägs S, Meyer S, Altmannshofer S, Opel M, Wilhelm F, Rogalev A, et al. Investigation of induced *Pt* magnetic polarization in *Pt*/$Y_3Fe_5O_{12}$ bilayers. Applied Physics Letters. 2012;**101**(26):262407. DOI: 10.1063/1.4773509

[31] Huang SY, Fan X, Qu D, Chen YP, Wang WG, Wu J, et al. Transport magnetic proximity effects in platinum. Physical Review Letters. 2012;**109**: 107204. DOI: 10.1103/PhysRevLett.109.107204

[32] Lu YM, Choi Y, Ortega CM, Cheng XM, Cai JW, Huang SY, et al. *Pt* magnetic polarization on $Y_3Fe_5O_{12}$ and magnetotransport characteristics. Physical Review Letters. 2013;**110**: 147207. DOI: 10.1103/PhysRevLett.110.147207

[33] Hirsch JE. Spin Hall effect. Physical Review Letters. 1999;**83**: 1834-1837. DOI: 10.1103/PhysRevLett.83.1834

[34] Edelstein V. Spin polarization of conduction electrons induced by electric current in two-dimensional asymmetric electron systems. Solid State Communications. 1990;**73**(3):233-235. Available from: https://www.sciencedirect.com/science/article/pii/003810989090963C

[35] Bel'kov VV, Ganichev SD. Magneto-gyrotropic effects in semiconductor quantum wells. Semiconductor Science and Technology. 2008;**23**(11):114003. DOI: 10.1088/0268-1242/23/11/114003

[36] Miron I, Garello K, et al. Perpendicular switching of a single ferromagnetic layer induced by in-plane current injection. Nature. 2011;**476**: 189-193

[37] Haldar A, Adeyeye AO. Functional magnetic waveguides for magnonics. Applied Physics Letters. 2021;**119**(6): 060501. DOI: 10.1063/5.0061528

[38] Sinova J, Culcer D, Niu Q, Sinitsyn NA, Jungwirth T, MacDonald AH. Universal intrinsic spin hall effect. Physical Review Letters. 2004;**92**:126603. DOI: 10.1103/PhysRevLett.92.126603

[39] Demidov VE, Demokritov SO, Rott K, Krzysteczko P, Reiss G. Linear and nonlinear spin-wave dynamics in macro- and microscopic magnetic confined structures. Journal of Physics D: Applied Physics. 2008;**41**(16): 164012. DOI: 10.1088/0022-3727/41/16/164012

[40] Yu H, Huber R, Schwarze T, Brandl F, Rapp T, Berberich P, et al. High propagating velocity of spin waves and temperature dependent damping in a CoFeB thin film. Applied Physics Letters. 2012;**100**(26):262412. DOI: 10.1063/1.4731273

[41] Sebastian T, Ohdaira Y, Kubota T, Pirro P, Brächer T, Vogt K, et al. Low damping spin-wave propagation in a micro-structured $Co_2Mn_{0.6}Fe_{0.4}Si$ heusler waveguide. Applied Physics Letters. 2012;**100**:112402. Available from: https://api.semanticscholar.org/CorpusID:121064753

[42] Lim J, Bang W, Trossman J, Kreisel A, Jungfleisch MB, Hoffmann A, et al. Direct detection of multiple backward volume modes in yttrium iron garnet at micron scale wavelengths. Physical Review B. 2019;**99**:014435. DOI: 10.1103/PhysRevB.99.014435

[43] Seavey MH, Tannenwald PE. Direct observation of spin-wave resonance. Physical Review Letters. 1958;**1**:168-169. DOI: 10.1103/PhysRevLett.1.168

[44] Eshbach JR. Spin-wave propagation and the magnetoelastic interaction in yttrium iron garnet. Physical Review Letters. 1962;**8**:357-359. DOI: 10.1103/PhysRevLett.8.357

[45] Rezende SM, Morgenthaler FR. Magnetoelastic waves in time-varying magnetic fields. I. Theory. Journal of Applied Physics. 1969;**40**(2): 524-536

[46] Rezende SM. Magnetoelastic and magnetostatic waves in time-varying magnetic fields [Ph.D. dissertation]. Massachusetts Institute of Technology; 1967. Available from: https://apps.dtic.mil/sti/citations/AD0664333

[47] Mahmoud A, Ciubotaru F, Vanderveken F, Chumak AV, Hamdioui S, Adelmann C, et al. Introduction to spin wave computing. Journal of Applied Physics. 2020; **128**(16):161101. DOI: 10.1063/5.0019328

[48] Moore G. Cramming more components onto integrated circuits. Proceedings of the IEEE. 1998;**86**(1): 82-85

[49] Wolf SA, Awschalom DD, Buhrman RA, Daughton JM, von Molnár S, Roukes ML, et al. Spintronics: A spin-based electronics vision for the future. Science. 2001;**294**(5546): 1488-1495. Available from: https://www.science.org/doi/abs/10.1126/science.1065389

[50] Sander D, Valenzuela SO, Makarov D, Marrows CH, Fullerton EE, Fischer P, et al. The 2017 magnetism roadmap. Journal of Physics D: Applied Physics. 2017;**50**(36):363001. DOI: 10.1088/1361-6463/aa81a1

[51] Bernstein K, Cavin RK, Porod W, Seabaugh A, Welser J. Device and architecture outlook for beyond cmos switches. Proceedings of the IEEE. 2010; **98**(12):2169-2184

[52] Žutić I, Fabian J, Das Sarma S. Spintronics: Fundamentals and applications. Reviews of Modern Physics. 2004;**76**:323-410. Available from: https://link.aps.org/doi/10.1103/RevModPhys.76.323

[53] Hall KC, Flatté ME. Performance of a spin-based insulated gate field effect transistor. Applied Physics Letters. 2006;**88**(16):162503. DOI: 10.1063/1.2192152

[54] Imre A, Csaba G, Ji L, Orlov A, Bernstein GH, Porod W. Majority logic gate for magnetic quantum-dot cellular automata. Science. 2006;**311**(5758): 205-208. Available from: https://www.science.org/doi/abs/10.1126/science.1120506

[55] Allwood DA, Xiong G, Faulkner CC, Atkinson D, Petit D, Cowburn RP. Magnetic domain-wall logic. Science. 2005;**309**(5741): 1688-1692. Available from: https://www.science.org/doi/abs/10.1126/science.1108813

[56] Nikonov DE, Bourianoff GI, Ghani T. Proposal of a spin torque majority gate logic. IEEE Electron Device Letters. 2011;**32**(8):1128-1130

[57] Koumpouras K, Yudin D, Adelmann C, Bergman A, Eriksson O, Pereiro M. A majority gate with chiral magnetic solitons. Journal of Physics: Condensed Matter. 2018;**30**(37):375801. DOI: 10.1088/1361-648X/aad82f

[58] Barman A, Gubbiotti G, Ladak S, et al. The 2021 magnonics roadmap. Journal of Physics: Condensed Matter. 2021;**33**(41):413001. DOI: 10.1088/1361-648X/abec1a

[59] Kostylev MP, Serga AA, Schneider T, Leven B, Hillebrands B. Spin-wave logical gates. Applied Physics Letters. 2005;**87**(15):153501. DOI: 10.1063/1.2089147

[60] Khitun A, Wang KL. Non-volatile magnonic logic circuits engineering. Journal of Applied Physics. 2011;**110**(3): 034306. DOI: 10.1063/1.3609062

[61] Chumak AV. Spintronics Handbook, Second Edition: Spin Transport and Magnetism. CRC Press; 2019. ch. Magnonic Spintronics. DOI: 10.1201/9780429441189

[62] Hellman F, Hoffmann A, Tserkovnyak Y, Beach G, et al. Interface-induced phenomena in magnetism. Reviews of Modern Physics. 2017;**89**: 025006. Available from: https://link.aps.org/doi/10.1103/RevModPhys.89.025006

[63] Ganguly A, Rowan-Robinson RM, Haldar A, Jaiswal S, Sinha J, Hindmarch AT, et al. Time-domain detection of current controlled magnetization damping in $Pt/Ni_{81}Fe_{19}$ bilayer and determination of Pt spin Hall angle. Applied Physics Letters. 2014; **105**(11):112409. DOI: 10.1063/1.4896277

[64] Panda SN, Mondal S, Sinha J, Choudhury S, Barman A. All-optical detection of interfacial spin transparency from spin pumping in *Ta/CoBeB* thin films. Science Advances. 2019;**5**(4): eaav7200 https://www.science.org/doi/abs/10.1126/sciadv.aav7200

[65] Chaurasiya AK, Banerjee C, Pan SSS, Choudhury S, Sinha ABJ. Direct observation of interfacial Dzyaloshinskii-Moriya interaction from asymmetric spin-wave propagation in $W/CoFeB/SiO_2$ heterostructures down to sub-nanometer *CoFeB* thickness. Scientific Reports. 2016;**6**(32592):1-8

[66] Cubukcu M, Boulle O, Mikuszeit N, Hamelin C, Brcher T, Lamard N, et al. Ultra-fast perpendicular spin–orbit torque MRAM. IEEE Transactions on Magnetics. 2018;**54**(4):1-4

[67] Aradhya SV, Rowlands GE, Oh J, Ralph DC, Buhrman RA. Nanosecond-timescale low energy switching of in-plane magnetic tunnel junctions through dynamic oersted-field-assisted spin hall effect. Nano Letters. 2016;**16**(10): 5987-5992

Chapter 5

Graphene-Based Optical Waveguides for Surface Plasmon Polariton Transmission

Mohammad Javad Maleki, Mohammad Soroosh and Fatemeh Haddadan

Abstract

In this chapter, various ideas and designs related to plasmonic waveguides are reviewed. As plasmonic structures can confine light in small sizes, they can be considered a suitable option in the design of optical devices. In recent years, graphene has been known as an attractive material in optical applications and can be used to guide surface plasmons. By photo-stimulating surface plasmons at the semiconductor-graphene interface, plasmon polaritons can be confined near graphene and guided to several micrometers. Using this feature, various optical waveguides can be designed. The real and imaginary parts of the effective refractive index of the waveguide, the figure of merit, coupling length, crosstalk, and loss in decibels per micrometer are the most important parameters in the design of plasmonic waveguides. In recent years, various ideas such as the use of ridges, graphene nano-ribbons or strips, and graphene sheets have been proposed, which provide waveguides with different characteristics. In this chapter, the theory and basic relationships in light confinement in plasmonic waveguides are investigated first. In the following, different structures designed in recent years are reviewed.

Keywords: confinement, graphene, optical switch, optical waveguide, plasmonic device, surface plasmon polariton

1. Introduction

With the advancement of technology toward the realization of integrated photonic circuits, plasmonic structures and plasmonic waves have been the focus of attention and utilization. Most of these nanostructures are made up of metals, graphene, and dielectrics, with sizes below the excitation wavelength, which is the wavelength of the incident radiation that excites the plasmonic waves. Plasmons are described based on the interaction between electromagnetic waves and conduction electrons in metals with nanosizes or graphene. Metals exhibit reflective behavior in the vicinity of the visible frequency spectrum and do not allow the propagation of electromagnetic waves inside the metal. Therefore, metals are used for waveguide structures and oscillators. In higher frequencies such as visible light frequency, the penetration of the field into the metal increases significantly, leading to severe losses and consequently limiting the fabrication of optical device structures. In ultraviolet frequencies, metals behave like insulators, and electromagnetic waves can penetrate inside them. All the scattering properties mentioned above are described by the permittivity function as a function of frequency [1].

2. Surface plasmon polariton

Surface plasmon polaritons (SPPs) are excitations or electromagnetic waves that couple to the collective and oscillating free electrons on the surface of metals to propagate longitudinally on the interface between the metal and dielectric [2]. However, the incident electromagnetic radiation should meet certain conditions, including being incident in a way that creates total internal reflection and having longitudinal polarization. **Figure 1** illustrates the excitation and propagation of plasmonic waves at the interface of two media.

The reason for the importance of SPPs is their ability to concentrate the electromagnetic field. As a result, the scattering of plasmons decreases, and the propagation of localized fields improves by several orders of magnitude [3]. Therefore, it causes transverse electric or transverse magnetic fields to be confined to the surfaces perpendicular to their propagation direction.

3. Graphene

Researchers have been able to improve certain desired features in devices using new materials with special electro-optic properties. One of the most important materials that can be used for sensing and providing unique characteristics is graphene. Graphene has become a unique material due to its exceptional properties in electrical and thermal conductivity, high density, carrier mobility, optical conductivity, and mechanical properties, which have been employed in optical and electronic devices alongside three-dimensional materials such as silicon [4]. In recent years, graphene plasmons have garnered significant attention from researchers. Graphene, in the plasmonic phase, has led to the proposal of new optical devices in the terahertz to visible frequency range with very high speeds, low excitation voltages, low power consumption, and reduced size [5–8].

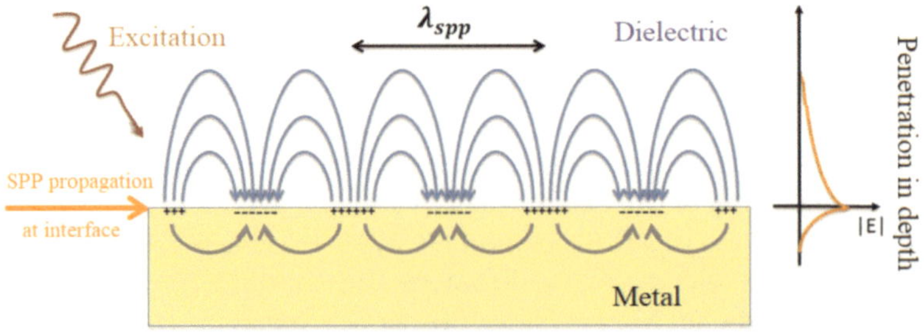

Figure 1.
Propagation of surface plasmon polariton waves at the interface of dielectric and metal and penetration of electric field into the two materials.

Surface plasmon waves are not radiative and appear exponentially at the boundary between two regions. This wave enables efficient coupling with input signals across a wide range of wavelengths while minimizing losses. Previously, noble metals like gold and silver were primarily used as plasmonic materials, allowing surface plasmon propagation from visible to infrared frequencies. However, these structures, particularly in the visible light range, exhibited significant losses and limited adjustability. Additionally, there are limitations on the ratio of the SPP propagation length at the metal-dielectric interface. One of the main reasons for using graphene is the possibility of tuning due to energy bandgap control, increased limitation due to graphene's thickness being comparable to an atom, and higher speed due to reduced electron scattering and fewer collisions [9]. Furthermore, in the optical frequency range, the incident light polarization has no significant effect on the graphene refractive index due to its high refractive index. For these reasons, graphene has been proposed as a suitable option for plasmonic structures in the terahertz frequency range. The reason for using graphene is that photons can easily couple with graphene surface plasmons in the terahertz frequency range, leading to remarkable features such as tunability with external voltage application and low losses.

Graphene is a two-dimensional crystal composed of carbon atoms arranged in a hexagonal lattice with sp2 bonding orbitals. In this structure, there is no energy gap between the valence and conduction bands, which have a symmetric Dirac-like cone shape and touch at the $k = 0$ point in the Brillouin zone. The possibility of conducting in this material is provided by engineering the energy gap through chemical doping or using a biased electric field or voltage. Using the voltage connection to graphene, the Fermi energy level can be changed. If a positive voltage is applied, the Fermi level increases, resulting in an increase in electron density and graphene acting as a p-type semiconductor [9].

The absence of the forbidden energies in the energy bands of graphene has led to unique optical, electronic, and mechanical properties, making it suitable for a wide range of optical applications. As a result, extensive research on graphene has been conducted by research centers and industries. A group of researchers successfully extracted graphene using a micro-mechanical approach from graphite material [10]. When we write with a pencil on paper, the remaining mark on the paper is graphite. If the applied pressure is adjusted so that only a one-atom-thick layer of graphite remains on the paper, graphene is produced. The graphene layer is then placed on silicon dioxide and compressed, transferring the graphene layer to the substrate. Although this method provides better quality compared to other methods, it is not proper for mass production [10]. Chemical synthesis is another common method for producing graphene. The first step is graphite oxidation using acids such as nitric acid and sulfuric acid to produce graphene oxide. Then, graphene oxide is converted to graphene using a thermal shock or ultrasonic waves.

Graphene is classified into three categories based on layer thickness or atomic layers: single-layer graphene, few-layer graphene (2–10 layers), and multilayer graphene (more than 10 layers). Graphene with more than 100 layers, or thick graphene, possesses graphite-like properties [11]. Atomic force microscopy is one of the methods used to determine the number of graphene layers [12]. Single-layer graphene absorbs about 2–3% of white light regardless of the size of the light's wavelength, which makes it suitable for optical devices such as photodetectors, solar cells, switches, and polarizers [13]. Due to its high mobility, controllable conductivity without carrier scattering, and increased penetration depth by a few micrometers, graphene is suitable for plasmonic structures.

The integrated structure of graphene allows for faster phonon transfer and increased heat transfer. The tunability of graphene is due to its mixed dynamic conductivity, which is determined by the Kubo formula. This formula indicates that dynamic conductivity depends on parameters such as chemical potential, scattering coefficient, and temperature. The concept of dynamic conductivity in graphene may vary in different frequency ranges depending on the positive or negative chemical potential. In intrinsic graphene or graphene with low chemical potential, the concept of dynamic conductivity is negative. In this state, graphene can transmit TE surface plasmon modes, acting as a semiconductor. By applying an external voltage, graphene has a large chemical potential; therefore, it can pass surface plasmon modes TM and act as a metal. Increasing temperature also increases the real part of dynamic conductivity, and when TM modes are conducting, the propagation constant in this state also increases [14, 15].

The conductivity of graphene is generally a complex quantity resulting from intraband and interband transitions [16]. In the visible frequency range, if the energy of photons is lower than the Fermi energy, the Fermi energy is centered around the Dirac point, and all the valence bands are filled and the conduction band is empty. There always exists a pair of excitable electron-holes with a corresponding photon energy, which results in effective photon absorption across all frequency ranges up to the visible range. When the Fermi level increases, lower energy photons below the Fermi level have a lower probability of absorption because of the Pauli exclusion. Therefore, in the visible frequency range, photons cause the transition of electrons from the valence band to the conduction band by creating electron-hole pairs. If excitation occurs, the photon energy is transferred to the electron. This type of transfer is called intraband transfer.

For calculating the conductivity factor of graphene in the visible frequency range, the intraband transition term is important, while the interband transition term is disregarded. In the terahertz frequency range, as there are impurities present in the medium, the Fermi energy is much higher than the photon energy. Consequently, according to the Pauli exclusion principle, intraband excitation in the terahertz frequency spectrum is negligible. Therefore, in terahertz frequencies, only interband transitions are significant [16].

The optical properties of graphene can be examined using the surface conductivity. The surface conductivity of graphene depends on various parameters, including frequency, temperature, chemical potential, and carrier mobility. By changing the surface conductivity of graphene, the optical properties of this two-dimensional material can be altered, which can be obtained from the Kubo formula and includes both intraband and interband contributions [17].

$$\sigma(\omega, T, \Gamma, \mu_c) = \sigma_{\text{intra}}(\omega, T, \Gamma, \mu_c) + \sigma_{\text{inter}}(\omega, T, \Gamma, \mu_c) \tag{1}$$

$$\sigma_{\text{intra}}(\omega, T, \Gamma, \mu_c) = \frac{-ie^2}{\pi\hbar^2(\omega + i2\Gamma)}\int_0^\infty \xi\left(\frac{\partial f_d(\xi)}{\partial(\xi)} - \frac{\partial f_d(-\xi)}{\partial(\xi)}\right)d\xi \tag{2}$$

$$\sigma_{\text{inter}}(\omega, T, \Gamma, \mu_c) = \frac{ie^2(\omega + i2\Gamma)}{\pi\hbar^2}\int_0^\infty \xi\left(\frac{f_d(-\xi)}{(\omega + i2\Gamma)^2} - \frac{f_d(\xi)}{4(\xi/\hbar)^2}\right)d\xi \tag{3}$$

where σ_{intra} is the intraband conductivity which represents absorption due to photon-electron scattering within the band. σ_{inter} refers to the interband conductivity and is related to electron transitions between two bands. The intraband conductivity is affected by temperature (T), carrier scattering rate (Γ), chemical potential (μ_c), angular frequency (ω), and reduced Planck constant ($\hbar$). It is described by the Fermi-Dirac distribution function (f_d) and can be obtained from Eq. (4) [18].

$$f_d = (1 + \exp((\varepsilon - \mu_c)/k_B T))^{-1} \tag{4}$$

By utilizing the above equations, the intraband conductivity of graphene can be rewritten as Eq. (5) [18].

$$\sigma_{intra}(\omega, T, \Gamma, \mu_c) = \frac{e^2 k_B T}{\pi \hbar^2 (2\Gamma + i\omega)} \left[\frac{\mu_c}{k_B T} + 2\ln\left(e^{-\mu_c/k_B T} + 1\right) \right] \tag{5}$$

At room temperature, the interband conductivity is obtained from Eq. (6) [19].

$$\sigma_{inter}(\omega, T, \Gamma, \mu_c) = -j\frac{e^2}{4\pi h^2} \ln\left(\frac{2|\mu_c| - (\omega - j2\Gamma)h}{2|\mu_c| + (\omega - j2\Gamma)h}\right) \tag{6}$$

For $\hbar\omega < 2\mu_c$, graphene conductivity is purely intraband, and the imaginary part of graphene's surface conductivity is positive, behaving like a metal and sustaining TM polarization. For $\hbar\omega > 2\mu_c$, graphene conductivity is interband, and the imaginary part of conductivity is negative, behaving like an insulator and supporting propagating TE polarization [20]. Therefore, in the average infrared wavelengths, graphene conductivity is defined as a semi-metallic conductivity model and only includes the intraband conductivity term, which is obtained from Eq. (7) at room temperature [19].

$$\sigma_g = \frac{ie^2 \mu_c}{\pi e \hbar^2 (\omega + i\tau^{-1})} \tag{7}$$

where τ is the carrier relaxation time and is obtained from Eq. (8) [19].

$$\tau = \frac{\upsilon \mu_c}{e V_F^2} \tag{8}$$

where υ is the carriers' mobility and its value is 10^4 cm^2/Vs. V_F represents the speed and equals 10^6 m/s. The chemical potential of graphene is adjustable by applying an external electric field or doping. The chemical potential is related to the bias voltage according to Eq. (9) [21].

$$\mu_c = \hbar V_F \sqrt{\frac{\pi \varepsilon_0 \varepsilon_r V_{bias}}{e D_1}} \tag{9}$$

where ε_r, D_1, and ε_0 are the relative permittivity, substrate layer thickness, and vacuum permittivity, respectively. The sheet conductivity of graphene is obtained from Eq. (10) [22].

$$\varepsilon_q = 1 + \frac{i\sigma_g}{\varepsilon_0 \omega t} \tag{10}$$

4. Theory of SPP excitation

A wave with TM polarization incident on a surface with an angle θ_1 is shown in **Figure 2**. The momentum of the photon in the dielectric equals $hk_d/2\pi$, where n_d is the refractive index of the dielectric and $k_d = 2\pi n_d/\lambda$.

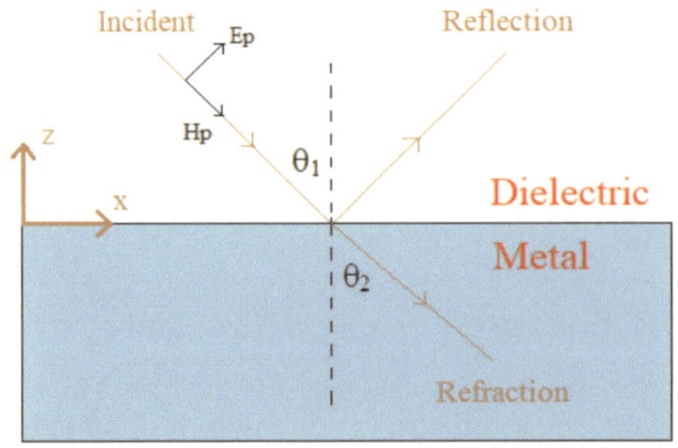

Figure 2.
The interface between a metal and a semiconductor, considering optical illumination with TM polarization.

When a wave illuminates the metal-dielectric interface, a portion of it at a similar angle is reflected, and its momentum is preserved in the dielectric. Additionally, a portion of the incident optical wave is transmitted through the metal at an angle of θ_2, carrying a momentum of $\hbar k_m$ in the metal. Wavevector k_m represents the metal's index of the metal through $k_m = 2\pi n_m/\lambda$. The component at the x-axis is conserved, meaning $k_m x = k_d x$, where $k_d x = k_d \sin\theta_1$ and $k_m x = k_m \sin\theta_2$. Thus, it can be written as [3]:

$$n_d \sin\theta_1 = n_m \sin\theta_2 \tag{11}$$

The above relation is known as Snell's law. Generally, in the visible wavelength range, n_d is greater than n_m. As $n_d > n_m$, θ_2 is maximized at 90 degrees, and θ_1 is limited. For angles greater than the limiting value, the incident wave is not transmitted into the metal. The limiting value of the incidence angle is known as the critical value θ_c, and it can be concluded that:

$$\sin\theta_c = n_m/n_d \tag{12}$$

A wave with $\theta_1 > \theta_c$ has a greater momentum along the interface compared to the metal. Incident-polarized P at the interface of the metal and dielectric possesses electric and magnetic fields that create oscillating surface charges on the common surface between them. Although the wave with an angle greater than the critical angle is completely reflected from the common surface, it causes the surface charges to undergo oscillatory motion and generate a field. This field penetrates into the metal, and these induced fields are perpendicular to the common surface.

When a wave illuminates the surface at the critical angle, there is an infinite extent of the surface where surface charges can oscillate. However, as the incident angle increases further, the oscillation of surface charges decreases and is emitted at a shorter distance from the surface. In these cases, the induced plasmon fields created due to the wave incident at an angle greater than the critical angle are considered as factors in coupling photons to surface plasmons.

There are no boundaries for E_x, so it is preserved at the overall surface. This is not true for E_z, which is the vertical component of the electric field. The vertical component D_z is continuous, and E_z changes with the displacement and the dielectric coefficient because $D_z = \varepsilon_d \varepsilon_0 E_z$ and $D_z = \varepsilon_m \varepsilon_0 E_{zm}$. Discontinuity in component E_z leads to polarization changes at the surface.

A normal incident wave with s polarization does not create surface charges at the interface, but a radiating wave with p polarization makes polarization charges. Due to the need for electric field components perpendicular to create surface charges, only electromagnetic waves with p polarization should be considered. Additionally, any surface wave should satisfy the electromagnetic wave equation in both dielectric and metal materials. Assuming the x-y plane for the metal-dielectric, one has the following formulas in the x direction for $z > 0$:

$$E_d = (E_{xd}, 0, E_{zd})\ \exp(-k_{zd}z)\ \exp[i(k_x x - \omega t)] \tag{13}$$

$$H_d = (0, H_{yd}, 0)\ \exp(-k_{zd}z)\ \exp[i(k_x x - \omega t)] \tag{14}$$

For $z < 0$;

$$E_m = (E_{xm}, 0, E_{zm})\ \exp(-k_{zm}z)\ \exp[i(k_x x - \omega t)] \tag{15}$$

$$H_m = (0, H_{ym}, 0)\ \exp(-k_{zm}z)\ \exp[i(k_x x - \omega t)] \tag{16}$$

Taking Eqs. (13) and (15) into account in Maxwell's equation $\nabla.E = 0$, the components of the electric field can be written as follows:

$$E_{zd} = i\frac{k_x}{k_{zd}}E_{xd} \tag{17}$$

$$E_{zm} = -i\ \frac{k_x}{k_{zm}}\ E_{xm} \tag{18}$$

Taking Eqs. (13) and (14) into account, Maxwell's equation can be applied as Eq. (19)

$$\nabla \times E = -\frac{1}{c}\frac{\partial H}{\partial t} \tag{19}$$

Furthermore, we have:

$$-k_{zd}\ E_{xd} - ik_x E_{zd} = ikH_{yd} \tag{20}$$

where k refers to the wave vector in the free space and equals ω/c. By considering Eqs. (15), (16), and (19), we have:

$$k_{zm}\ E_{xm} - i\ k_x\ E_{zm} = i\ k\ H_{ym} \tag{21}$$

With regard to Eqs. (17), (18), (20) and (21) can be expressed as follows:

$$\varepsilon_d k E_{xd} = ik_{zd}H_{yd} \tag{22}$$

$$\varepsilon_m k E_{xm} = -ik_{zm}H_{ym} \tag{23}$$

In a way that:

$$k_{zd}^2 = k_x^2 - \varepsilon_d k^2 \tag{24}$$

$$k_{zm}^2 = k_x^2 - \varepsilon_m k^2 \tag{25}$$

Furthermore, the tangential components E and H are continuously related at the z = 0 boundary condition, meaning $E_{xd} = E_{xm}$ and $H_{yd} = H_{ym}$. The relationship between ε and the perpendicular k components in the metal and dielectric is expressed in Eqs. (4)–(16).

$$\frac{k_{zd}}{k_{zm}} = -\frac{\varepsilon_d}{\varepsilon_m} \tag{26}$$

Based on Eqs. (24), (25), and (26), one can conclude:

$$k_x = k\sqrt{\frac{\varepsilon_d \varepsilon_m}{\varepsilon_d + \varepsilon_m}} \tag{27}$$

If we replace k_x with k_{spp}, the scattering relationship SPP can be expressed as follows:

$$k_{spp} = k\sqrt{\frac{\varepsilon_d \varepsilon_m}{\varepsilon_d + \varepsilon_m}} \tag{28}$$

The momentum of SPP ($\hbar k_{spp}$) is greater than the momentum of a photon in free space ($\hbar k$) for the same frequencies, which causes the mismatch between light and SPP. It is necessary to overcome the challenge of the mismatch between light and SPP at the common interface. Therefore, it can be written as follows:

$$\varepsilon_d + \varepsilon_m = 0 \tag{29}$$

The metal permittivity, ε_m, is similar to that of free electrons.

$$\varepsilon_m = 1 - \frac{\omega_p^2}{\omega^2} \tag{30}$$

where ω_p is the plasma frequency, and for many metals, it falls in the ultraviolet wavelength region. Based on Eqs. (29) and (30), the SPP frequency can be written as:

$$\omega_{spp} = \frac{\omega_p}{\sqrt{1 + \varepsilon_d}} \tag{31}$$

5. Graphene-based plasmonic waveguides

This section represents a brief literature review of graphene-based plasmonic waveguides. Electrical conductivity in graphene is tuned by applying a bias voltage or doping, so different optical analog and digital devices can be designed using the graphene layer. Plasmonic waveguides due to their applications in communications have great attention.

5.1 Graphene on silver

Bagheri et al. proposed a plasmonic waveguide that features a graphene monolayer on silver, positioned between two SiO_2 layers [22]. The SiO_2 and silver layers have thicknesses of 200 and 30 nm, respectively. **Figure 3** depicts the transmission loss in the plasmonic waveguides across varying waveguide widths at chemical potentials of 0.4, 0.6, 0.8, and 1 eV. Increasing the chemical potential results in an elevation of the Fermi level and a reduction in the probability of occupation of energy levels exceeding the Fermi level. This phenomenon leads to a minor absorption in graphene, with loss diminishing as the chemical potential increases. Consequently, the modulation of transmittance is achievable by adjusting the chemical potential value. The high absorption exhibited at μ_c = 0.4 eV classifies graphene as a light absorber, which is unsuitable for this investigation. Enhancing transmittance is crucial for guiding terahertz waves in both the waveguides and the resonant ring. Therefore, a chemical potential of 1 eV is adopted in this analysis. As illustrated in **Figure 3**, an increase in waveguide width from 10 to 100 nm corresponds to a monotonic reduction in loss. The simulations indicate that field concentration at the waveguides' edges is significant for widths narrower than 30 nm. Considering the fundamental needs of optical circuits, a small area is recognized as a pivotal factor in devising optical devices. Hence, a waveguide width of 50 nm is chosen.

5.2 Parallel graphene with spacing

The first study in the field of parallel graphene plasmonic waveguides was conducted by Hanson [23]. In this research, the Maxwell equation for the propagation of quasi-transverse electromagnetic (TEM) mode in a parallel graphene-layer-based waveguide was solved. He considered a fundamental structure including three layers with the permittivity of ε_1, ε_2, and ε_3 where the graphene layer was at ε_1-ε_2 and ε_2-ε_3 interfaces. He calculated the mixed wavevector diagram β/k_0 and the surface conductivity of graphene σ/σ_{min} as a function of frequency for both the monolayer and 10-layer cases. The increase in β/k_0 frequency for both the monolayer and 10 layers converges to an approximately constant value. The imaginary part of β/k_0 tends toward zero due to the Pauli blocking principle. The behavior of the real part σ/σ_{min} for monolayer graphene shows a decreasing trend with frequency, while the imaginary part σ/σ_{min} exhibits an initial decrease followed by an increase.

Table 1 depicts the attenuation or loss coefficient (α) for monolayer graphene at a frequency of 100 GHz and distances of 100 nm for various values of chemical potential. The attenuation decreases with increasing chemical potential of graphene and, likewise, with the spacing among layers.

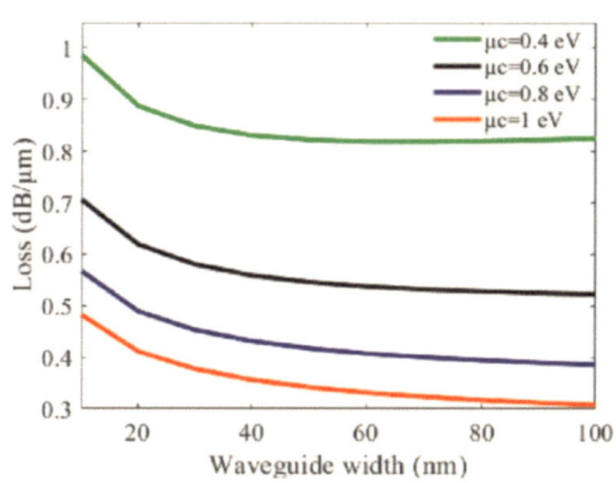

Figure 3.
Transmission loss in terms of waveguide width for different values of the graphene chemical potential.

μ_c (eV)	α (dB/μm)	
	d = 0.1 μm	d = 1 μm
0	1.22	0.394
0.4	0.362	0.115
0.8	0.256	0.081
1.2	0.209	0.066
1.6	0.198	0.061
2	0.162	0.05

Table 1.
Attenuation of monolayer graphene wave for various values of chemical potential.

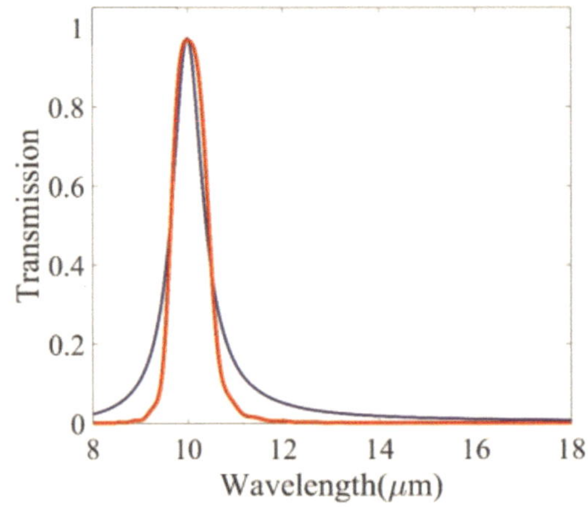

Figure 4.
The transmission spectrum for graphene tapes with a width of 20 nm and air gaps [24].

5.3 Graphene tape on SiO_2

Mohammadi et al. introduced a graphene monolayer 20 nm in width situated on the top of a silicon dioxide layer that is 50 nm thick, for the purpose of confining and guiding terahertz waves [24]. The configuration is subject to illumination in the transverse magnetic mode. As illustrated in **Figure 4**, the waves can pass through and be filtered using an air gap at a graphene chemical potential of 0.9 eV. The transmitted wave spectrum is concentrated around a central wavelength of 10 μm. The real component of the effective refractive index is approximately 80, indicating the effective confinement of surface plasmon polaritons. The imaginary part of the effective refractive index is 0.038, a suitable value for SPP transmission over short distances of a few micrometers.

5.4 Graphene-SiO_2-Si

He and Li conducted a comparison between surface plasmon modes in the transverse electric field (TE) and transverse magnetic field (TM) configurations for a graphene-SiO_2-Si structure [25]. A plasmonic channel was formed below the graphene in the silicon dioxide layer.

In this study, dispersion relations for surface plasmon modes in TE and TM configurations were extracted using propagation equations and boundary conditions. The effective refractive index and transmission loss for TM and TE modes in the graphene-SiO_2-Si structure with various SiO_2 layer thicknesses (50, 100, 200, and 300 nm) were calculated. The chemical potential of graphene was set to 0.2 eV. Calculations indicate that the TM mode is confined, and the refractive index of graphene in the TM mode increases with frequency. Consequently, the TM mode exhibits better confinement compared to metal-based structures. The maximum loss occurs at a frequency range of 2.5–3 THz. For frequencies higher than 4 THz, the effect of silicon dioxide thickness on the loss factor is negligible.

5.5 Graphene on gold

Ono et al. reported the use of graphene plasmonic waveguides with sizes smaller than 230 × 20 nm [26]. In this thin-layered waveguide, a gold layer is deposited on a SiO_2 substrate, and on top of it, a single layer of graphene is placed. The waveguide particularly confines light for a chemical potential of 0.2 eV. In this structure, graphene nonlinear absorption has increased, resulting in fast switching. The switching energy is 35 fJ, and the response time is 260 fs. The energy of switching is lower than conventional devices, and the switching time is the smallest reported to date. This device can effectively integrate with common silicon waveguides used in photonic integrated circuits. They calculated the field enhancement factor for changes in the thickness of gold from 20 to 100 nm. The results demonstrated the thinner gold layer makes the higher enhancement.

5.6 Multilayers of graphene and dielectric

A graphene-based plasmonic waveguide was introduced [27]. This waveguide utilized a multilayer graphene and dielectric structure positioned on the top of a silicon waveguide core, resulting in reduced loss and significantly increased propagation length.

According to the reported results, an increase in the number of dielectric layers leads to a reduction in the effective refractive index of the waveguide, diminishing light confinement. Simultaneously, the rise in the number of dielectric layers results in an extension of the propagation length, reaching approximately 30 μm. As a result, losses have been mitigated with the increase of graphene layers.

5.7 Ridge-based waveguide

Haddadan and Soroosh proposed a plasmonic waveguide [28]. This waveguide consists of a silicon ridge with a height (h) of 50 nm and a width (W) of 450 nm placed on silicon substrates with a thickness of 200 nm. On this ridge, a single-layer graphene is sandwiched between layers of silicon dioxide, with thicknesses of 20 and 12 nm. The space between silicon and graphene serves as the plasmonic waveguide or channel, aiding in trapping the incoming light in the waveguide and propagating it along the length of the waveguide (as shown in **Figure 5**).

The graphene chemical potential makes a difference in the refractive index of the silicon stack and its surrounding area. Therefore, due to the significant increase in the

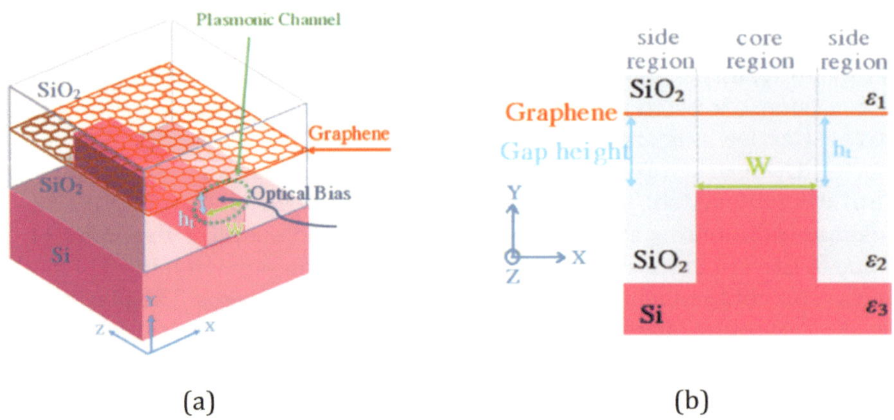

Figure 5.
(a) A view of the structure and (b) a representation in the x-y plane for a better understanding of the ridge-based channel [28].

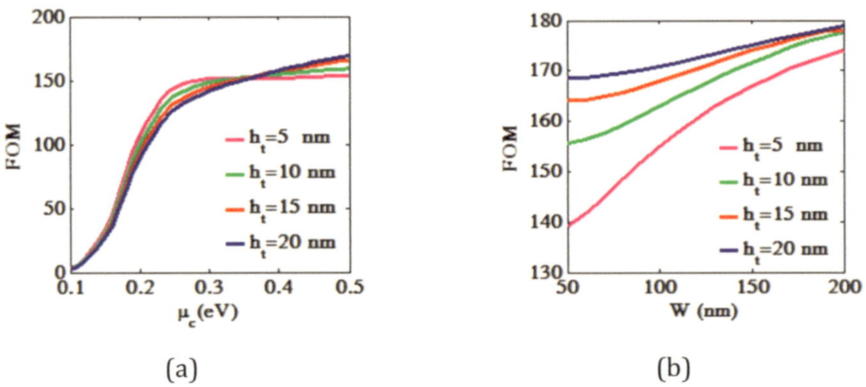

Figure 6.
FOM factor in terms of (a) chemical potential and (b) waveguide width for various channel heights [28].

electric field, light is focused in the plasmonic channel. This phenomenon is attributed to the slot effect. The slot confines terahertz waves in a region with a lower index due to the overall reflection trapping of light.

The analysis indicates that the height and width of the channel have a direct impact on FOM in the plasmonic waveguide. Increasing the width brings the behavior of the waveguide closer to a slab behavior, while increasing the height leads to an increase in FOM due to reduced loss. Additionally, raising the chemical potential and confinement results in an increased FOM, while the corresponding graphs for effective refractive index N and loss α show a decrease. These findings highlight the enhancement of light guidance performance in plasmonic channels through increased height and reduced loss (as depicted in **Figure 6**).

5.8 SiO_2-graphene-ridge

In another research, a plasmonic waveguide was designed by Maleki et al. [5]. As shown in **Figure 7a**, with the help of a ridge and adjustment of chemical potential, cross-talk reduces in comparison with the waveguide of Ref. [28].

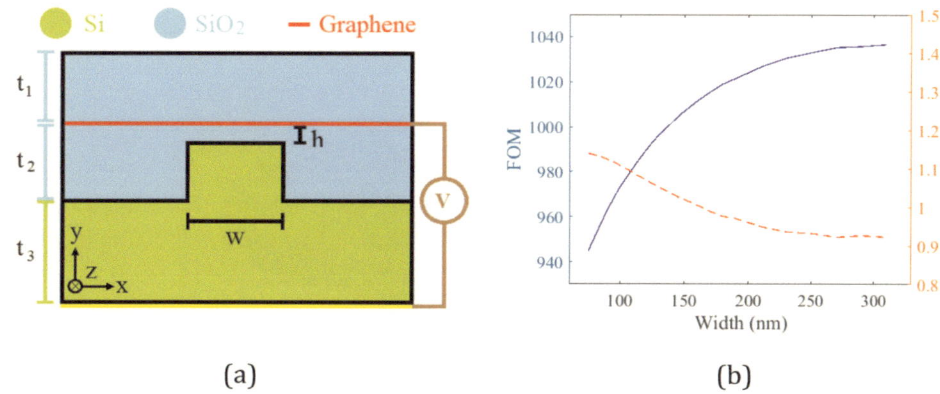

Figure 7.
(a) A view of the device in the cross-section and (b) FOM and loss in terms of the channel width in the x direction [5].

In **Figure 7b**, the graph illustrates the relationship between the channel width and both figures of merit and loss. The figure of merit ranges between 944.79 and 1036.63 as the width increases from 75 to 310 nm, while the loss decreases from 1.14 to 0.92 dB/μm within this width range. The superior figure of merit and minimal loss validate the exceptional performance of the channel under consideration. Hence, a width of 80 nm is selected for designing a compact waveguide, resulting in a figure of merit of 950.92 and a loss of 1.14 dB/μm.

5.9 SiO_2-graphene-SiO_2

A plasmonic structure utilizing graphene was suggested to guide and manipulate surface plasmon polaritons (as shown in **Figure 8**) [8]. The design involved positioning a graphene nanoribbon between two silicon dioxide layers. By inducing a 1 eV chemical potential to the graphene, a plasmonic waveguide formed close to the graphene monolayer. The effective confinement of the waveguide led to a decrease in channel loss to 0.029 dB/μm. Additionally, by reducing field penetration into the surrounding areas, the coupling length was extended to 187.9 μm.

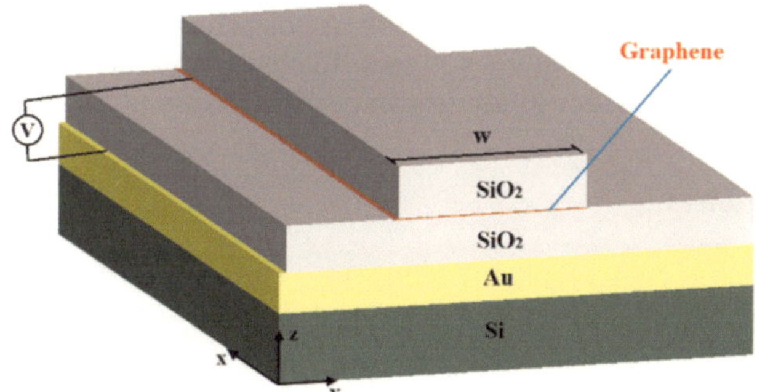

Figure 8.
A graphene monolayer between two silicon dioxide layers for the confinement of surface plasmon polaritons [8].

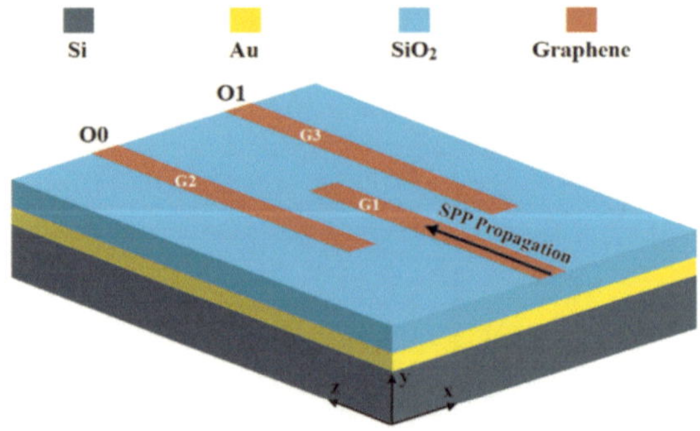

Figure 9.
Graphene nano-ribbon on silicon dioxide with a width of 50 nm for SPP confinement below the graphene [6].

5.10 Graphene nano-ribbon on SiO_2

Maleki et al. have designed a plasmonic channel with significant confinement to guide surface plasmon polaritons, utilizing 50 nm wide graphene nano-ribbons on a 230 nm thick silicon dioxide layer (as illustrated in **Figure 9**) [6]. A gold layer, with a thickness of 180 nm, was placed beneath the silicon dioxide to create a potential difference with respect to the graphene. The simulation results show that, by applying voltages of 1.5 and 8.3 V to graphene nano-ribbons, chemical potentials of 0.1 and 0.5 eV can be obtained, and the channel loss can be changed from 0.91 to 88.23 dB/μm. Based on this, two logical zero and one states and switching operations can be realized.

6. Conclusion

This chapter considers the theory of exciting surface plasmon polaritons and introduces the excellent features of graphene in the fields of optics and electricity. It also explores various concepts and structures designed to confine surface plasmon polaritons using graphene. The ridge-base structure, graphene nano-ribbon, and graphene sheet are highlighted as important methods for designing plasmonic devices. The adjustable chemical potential of graphene offers a promising avenue for manipulating terahertz wave transmission. The dependence of transmission loss on the chemical potential is extensively utilized in designing optical fast switches for both digital and analog applications. Through the investigation of relevant articles, it is demonstrated that a loss of 0.21 dB/μm and a figure of merit of up to 845 can be achieved for wavelengths ranging from 25 THz to 40 THz. These studies suggest that the graphene-dielectric interface holds considerable potential for guiding surface plasmons in plasmonic devices.

Acknowledgements

This work was supported by Shahid Chamran University of Ahvaz, grant number SCU.EE1402.672.

Conflict of interest

The authors declare no conflict of interest.

Author details

Mohammad Javad Maleki*, Mohammad Soroosh and Fatemeh Haddadan
Department of Electrical Engineering, Shahid Chamran University of Ahvaz, Ahvaz, Iran

*Address all correspondence to: mj.maleki@scu.ac.ir

References

[1] Sarid D, Challener WA. Modern Introduction to Surface Plasmons. UK: Cambridge University Press; 2013. DOI: 10.1017/CBO9781139194846

[2] Gric T. Plasmonics. Vol. 1. London, UK: Intechopen; 2018. pp. 14-18. DOI: 10.5772/intechopen.73373

[3] Zhang J, Zhang L, Xu W. Surface plasmon polaritons: Physics and applications. Journal of Physics D: Applied Physics. 2012;**45**:113001. DOI: 10.1088/0022-3727/45/11/113001

[4] Soroosh M, Farmani A, Maleki MJ, Haddadan F, Mansouri M. Highly efficient graphene-based optical components for networking applications. In: Photonic Crystal and its Applications for Next Generation Systems. Singapore: Springer Nature; 2023. pp. 15-35. DOI: 10.1007/978-981-99-2548-3_2

[5] Maleki MJ, Soroosh M, Akbarizadeh G. A subwavelength graphene surface plasmon polariton-based decoder. Diamond and Related Materials. 2023;**134**:109780. DOI: 10.1016/j. diamond.2023.109780

[6] Maleki MJ, Soroosh M, Akbarizadeh G. Design and simulation of compact plasmonic decoder with high contrast ratio to propagate graphene surface plasmon polaritons. Biquarterly Journal of Optoelectronic. 2024;**6**: 11-18. DOI: 10.30473/jphys.2023. 69675.1173

[7] Maleki MJ, Soroosh M, Akbarizadeh G. Design and simulation of compact subwavelength electro-optic switch based on graphene surface plasmon polaritons. Biquarterly Journal of Optoelectronic. 2023;**6**:41-50. DOI: 10.30473/jphys.2023.69371.1168

[8] Maleki MJ, Soroosh M. A low-loss subwavelength plasmonic waveguide for surface plasmon polariton transmission in optical circuits. Optical and Quantum Electronics. 2023;**55**:1266. DOI: 10.1007/ s11082-023-05603-0

[9] Zhu H, Xu Z, Xie D, Fang Y. Graphene Fabrication, Characterizations, Properties and Applications. India: Academic Press; 2018. DOI: 10.1016/C2016-0-02040-5

[10] Novoselov KS, Geim AK, Morozov SV, Jiang DE, Zhang Y, Dubonos SV, et al. Electric field effect in atomically thin carbon films. Science. 2004;**306**:666-669. DOI: 10.1126/ science.1102896

[11] Novoselov KS, Fal'ko VI, Colombo L, Gellert PR, Schwab MG, Kim K. A roadmap for graphene. Nature. 2012; **490**:192-200. DOI: 10.1038/nature11458

[12] Nemes-Incze P, Osváth Z, Kamarás K, Biró LP. Anomalies in thickness measurements of graphene and few layer graphite crystals by tapping mode atomic force microscopy. Carbon. 2008;**46**:1435-1442. DOI: 10.1016/j.carbon.2008.06.022

[13] Lashgari H, Abolhassani MR, Boochani A, Elahi SM, Khodadadi J. Electronic and optical properties of 2D graphene-like compounds titanium carbides and nitrides: DFT calculations. Solid State Communications. 2014; **195**:61-69. DOI: 10.1016/j.ssc.2014. 06.008

[14] Gusynin VP, Sharapov SG, Carbotte JP. Unusual microwave response of Dirac quasiparticles in graphene. Physical Review Letters. 2006; **96**:256802. DOI: 10.1103/ PhysRevLett.96.256802

[15] Li Z, Henriksen EA, Jiang Z, Hao Z, Martin MC, Kim P, et al. Dirac charge dynamics in graphene by infrared spectroscopy. Nature Physics. 2008;**4**: 532-535. DOI: 10.1038/nphys989

[16] Wang F, Zhang Y, Tian C, Girit C, Zettl A, Crommie M, et al. Gate-variable optical transitions in graphene. Science. 2008;**320**:206-209. DOI: 10.1126/science.1152793

[17] Chen J, Badioli M, Alonso-González P, Thongrattanasiri S, Huth F, Osmond J, et al. Optical nano-imaging of gate-tunable graphene plasmons. Nature. 2012;**487**:77-81. DOI: 10.1038/nature11254

[18] Wang DW, Zhao WS, Gu XQ, Chen W, Yin WY. Wideband modeling of graphene-based structures at different temperatures using hybrid FDTD method. IEEE Transactions on Nanotechnology. 2015;**14**:250-258. DOI: 10.1109/TNANO.2014.2387576

[19] Wang B, Zhang X, García-Vidal FJ, Yuan X, Teng J. Strong coupling of surface plasmon polaritons in monolayer graphene sheet arrays. Physical Review Letters. 2012;**109**:073901. DOI: 10.1103/PhysRevLett.109.073901

[20] Haddadan F, Soroosh M, Alaei-Sheini N. Cross-talk reduction in a graphene-based ultra-compact plasmonic encoder using an Au nano-ridge on a silicon substrate. Applied Optics. 2022;**61**:3209-3217. DOI: 10.1364/AO.449123

[21] Gómez-Díaz JS, Perruisseau-Carrier J. Graphene-based plasmonic switches at near infrared frequencies. Optics Express. 2013;**21**:15490-15504. DOI: 10.1364/OE.21.015490

[22] Bagheri F, Soroosh M. Design and simulation of compact graphene-based plasmonic flip-flop using a resonant ring. Diamond and Related Materials. 2023;**136**:109904. DOI: 10.1016/j.diamond.2023.109904

[23] Hanson GW. Quasi-transverse electromagnetic modes supported by a graphene parallel-plate waveguide. Journal of Applied Physics. 2008;**104**: 084314. DOI: 10.1063/1.3005881

[24] Mohammadi M, Soroosh M, Farmani A, Ajabi S. High-performance plasmonic graphene-based multiplexer/demultiplexer. Diamond and Related Materials. 2023;**139**:110365. DOI: 10.1016/j.diamond.2023.110365

[25] He XY, Li R. Comparison of graphene-based transverse magnetic and electric surface plasmon modes. IEEE Journal of Selected Topics in Quantum Electronics. 2014;**20**:62-67. DOI: 10.1109/JSTQE.2013.2257991

[26] Ono M, Hata M, Tsunekawa M, Nozaki K, Sumikura H, Chiba H, et al. Ultrafast and energy- efficient all-optical switching with graphene - loaded deep - subwavelength plasmonic waveguides. Nature Photonics. 2020;**14**:37-43. DOI: 10.48550/arXiv.1907.01764

[27] Huang CC, Chang RJ, Cheng CW. Ultra-low-loss mid-infrared plasmonic waveguides based on multilayer graphene metamaterials. Nanomaterials. 2021;**11**:2981. DOI: 10.3390/nano11112981

[28] Haddadan F, Soroosh M. Design and simulation of a subwavelength 4-to-2 graphene-based plasmonic priority encoder. Optics & Laser Technology. 2023;**157**:108680. DOI: 10.1016/j.optlastec.2022.108680

Chapter 6

Perspective Chapter: Optical Waveguides for Quantum Computation

Shiva Hayati Raad

Abstract

Quantum computation demands various types of optical devices such as beam splitters, ring resonators, delay lines, switches, modulators, multiplexers, Mach-Zehnder interferometers, and phase shifters, to name a few. These devices are essentially engineered optical waveguides, in which the geometrical and material properties are chosen such that the optical losses are minimized, to preserve the single photon's operation. Scattering loss, arising from the wall roughness, is the dominant loss mechanism in the optical waveguides, and its minimization should be taken into account, by either design or fabrication considerations. Different types of waveguide geometries including slab waveguides, strip waveguides, rib waveguides, Bragg grating waveguides, and hybrid waveguides have been utilized in this regard. Moreover, there are three main material categories, with generic fabrication processes, regarding the refractive index contrast of the core and cover. These include low, medium, and high index contrast materials, each owing its pros and cons. Finally, designing the bend waveguides with low bending loss is highly desirable, and circular bends, sinusoidal bends, Euler bends (normal and modified), and spline bends are some of the frequently used curvatures. This chapter reviews the key points in the design of optical waveguides for quantum applications.

Keywords: integrated photonics, index contrast, optical waveguide, quantum computation, bending loss, Euler bend

1. Introduction

Quantum calculation has been the subject of intensive research recently. In this technology, a single photon is exploited as the excitation source. Thus, it is essential to carefully control potential losses in the devices to reduce the probability of losing the single photon. Importantly, losses are the main source of error in quantum computation [1]. The total loss of a planar waveguide is the sum of the contributions of material absorption, Rayleigh scattering, interfacial scattering, substrate leakage, bend radiation, and cross-talk. The material losses in the photonic integrated circuits mainly originate from the bond resonances involving hydrogen atoms and can be attacked by high-temperature annealing. Also, the leakage losses can be prevented by the deposition of thick substrates, at the expense of fabrication time. The cross-talk can be attacked by considering sufficient distance between the waveguide cores. The interfacial losses change quadratically with the refractive index contrast and can be controlled either by material selection or fabrication processes or by introducing some specific processes, such as an oxidation-based etchless process, chemical oxidation, anisotropic etching, and post-processing [2, 3].

In classic computation, the processing speed is proportional to the number of transistors. According to the widespread rendering of Moor's law, the number of transistors on a single integrated circuit chip doubles every 18 months and the gradual miniaturization of electrical circuits is limited by quantum mechanical effects. Quantum computation, on the other hand, utilizes the superposition phenomenon, where the state of the quantum bits (qubits) is the superposition of the zero and one states. This feature increases the processing speed dramatically, since in many-qubit systems, the superposition state increases with the number of qubits [4, 5]. The electronic devices perform based on fermions (electrons), which obey Fermi–Dirac statistics, whereas, on the photonic devices, bosons (photons) are employed, obeying Bose–Einstein statistics [6].

Regarding the technology, quantum computation can be performed by superconducting qubits that demand dilution refrigerators to maintain their temperatures sufficiently low [7]. Moreover, the bulk and low-loss fiber optic system, as the name implies, is bulky, and reaching a stable performance is challenging [2, 8]. The integrated optic is highly desirable because of substantial upgradation in performance, reliability, and SWaP-C (size, weight, power, and cost). This is due to the integration of components, reduced coupling losses, and reduced sensitivity to environmental perturbation [9]. The photonic integrated circuits rely on processing the light confined to the waveguides [10]. In this technology, the quantum information is essentially the quantum states of light propagating in the photonic integrated circuit [11].

Integrated photonic platforms are formed by three main parts: a single-photon source (SPS), a circuit, and a single-photon detector (SPD). These sections are respectively utilized for generating, manipulating, and detecting the light on a small chip. Parametric photon-pair source and quantum dot (QD) single-photon source are two instances of SPSs. Moreover, avalanche photodiodes, superconducting nanowire single-photon detectors (SNSPDs), and transition edge sensors (TESs) are some of the SPD technologies [12, 13]. Similar to the classic gates, there are basic quantum gates including X, Y, Z, and Hadamard. Quantum gates are represented by unitary matrices, and each mathematical calculation can be separated into multiple such unitary calculations [14].

The ability to control the light propagation in the device after fabrication is a great demand in quantum calculation. The reconfigurable performance of the waveguides is implemented *via* (1) temperature control, (2) applying mechanical strain, (3) applying a magnetic field, (4) applying an electric field to lithium niobate ($LiNbO_3$) and 2-D materials, and (5) free-carrier plasma effect [15–18]. These methods modify the local refractive index of the waveguide modes, thus providing a phase shift in the performance. The tuning capability of the devices can be used to compensate for the impact of the fabrication imperfections in the spectral response as well [19].

Among the aforementioned methods, the thermo-optic effect offers a large dynamic range, moderate power consumption, negligible optical loss, and moderate operating speed (a few to tens of microseconds reconfiguration time) [20]. The challenge of the thermo-optic effect is the difficulty of local temperature rise in a tiny structure, which can be obviated by injecting charge carriers, providing much faster tuning time than thermal tuning method in the order of nanosecond [16, 21]. The graphene-based heaters may provide higher response speeds than those of the metallic ones [22].

2. Straight and bend waveguide design

This section discusses the various straight waveguide geometries and the key points for designing each configuration. The geometrical dimension selection and the choice of the platform play a crucial role in this regard. To extend the waveguide design to an integrated device design, it is essential to use engineered curved waveguides for connecting different sections. The main goal of the bend waveguide design is the reduction of the losses and curvatures with different mathematical formulas can be used. These materials are illustrated in the second part of this section.

2.1 Straight waveguide geometries

Integrated photonics relies on the design of simple waveguides, where other components are essentially built by introducing geometrical or physical modification to the base waveguide [23]. The performance of the waveguides is mainly evaluated by their propagation constant, attenuation constant, and insertion loss. Moreover, the transverse profile of the dielectric constant of the waveguide is employed to categorize the modes. Waveguide modes are denoted by transverse electric (TE) and transverse magnetic (TM). Moreover, degenerate modes share the same value of propagation constant but with distinguishable electric field distributions [24].

Different types of full-wave solvers have been used to simulate the performance of the optical waveguides, including but not limited to Ansys Lumerical, Comsol Multiphysics, FIMMWAVE, Rsoft, and CST commercial packages [25–29]. These solvers are based on different numerical techniques, including the vectorial beam propagation method (BPM), the finite-difference time-domain (FDTD) method, the eigenmode expansion method (EME), and the finite-element mode propagation analysis (FE-MPA). For the initial design, usually 2D or 2.5D solvers are employed. The 3D implementation of the numerical methods leads to more accurate results at the expense of more time and memory requirements. Furthermore, the 3D calculation demands considering multiple parameters such as port widths and locations [30, 31].

The simplest form of the optical waveguide is the slab waveguide, which is a thin film core, with the refractive index n_1, sandwiched between upper and lower cladding, as shown in **Figure 1a**. The refractive index of the core is higher than those of the upper and lower claddings, respectively denoted by n_2 and n_3. This structure performs based on total internal reflection, and it shows a confined wave field in the core region and exponentially decaying wave in the clad. This boundary value problem has a closed-form solution, and the effective refractive index of each propagating mode lies between those of the core and cladding. If the effective mode index of the core is less than that of the clad, the wave in the clad becomes oscillatory in the transverse direction and dissipates as the radiating mode [33]. In this structure, the light is confined in only one direction, and high-power waveguide lasers and amplifiers benefit from this geometry [24].

The three-dimensional version of the slab waveguides is of the strip type, illustrated in **Figure 1b**. It is constructed by a high-index core on top of the substrate. The effective refractive indexes of these layers are denoted by n_1 and n_2. To reduce the leakage from the substrate, the core can be covered by a superstrate and form a waveguide, the so-called buried channel waveguide, as shown in **Figure 1c**. The width w and height d of the waveguide can be manipulated to excite the propagation of the desired modes. The three-dimensional waveguide can be formed by either a rectangular or cylindrical core, supported by a substrate and/or superstrate.

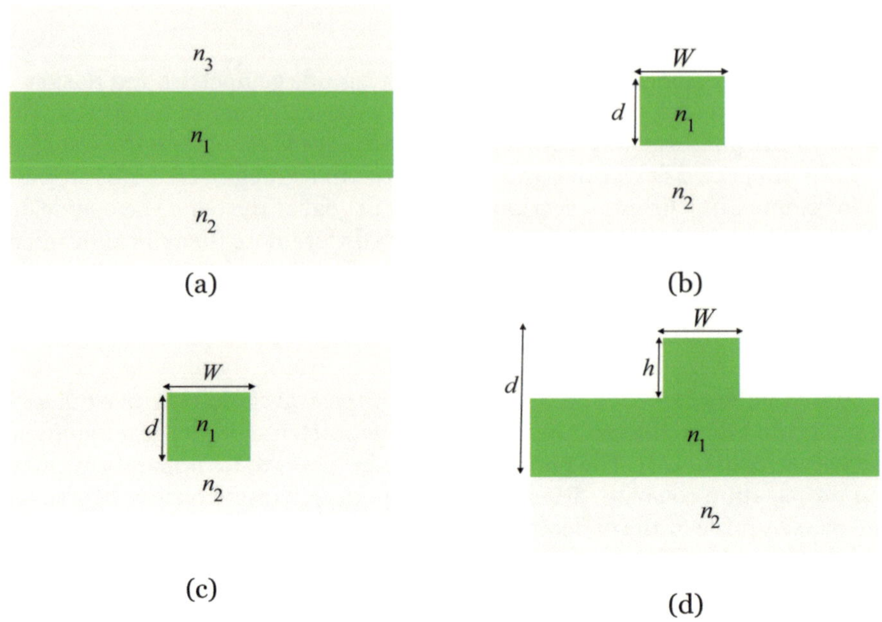

Figure 1.
The cross section of different waveguide geometries: (a) slab waveguide, (b) strip (ridge) waveguide, (c) buried channel waveguide, and (d) rib waveguide ($n_1 > n_2 > n_3$). The three-dimensional views of the waveguides can be found in Ref. [32].

The strip waveguide results in light confinement in two dimensions; thus, it is suitable for applications where diffraction-less performance is required. The device can be modeled by semi-analytical or numerical calculation since there is not any general exact analytical solution for this geometry and under specific conditions, approximate solutions can be attained [10]. Effective-index method, Marcatili's method, the mode-matching method, the finite-element method, the finite-difference method, the weighted-index method, and the perturbation theory are some of the implemented numerical tools for the analysis of these waveguides [34]. This geometry can be considered as the main building block for the optical device design.

To realize the propagating modes in optical waveguides, the mode dispersion diagram is used. This diagram shows the supported modes in the waveguide as a function of the waveguide width, and it is an essential tool for waveguide design [35]. A typical mode dispersion diagram is shown in **Figure 2a**, from which the onset of the propagation of each transverse mode can be recognized. Thus, it is feasible to propagate the desired number of modes in the waveguides by the proper choice of the core width for a given substrate. The electric field distribution for each transverse mode is shown in **Figure 2b** [36]. The anti-crossing between the TE and TM modes in the mode dispersion diagram is the evidence of hybridization, which can be further confirmed by the polarization ratio [37]. Such hybridization is not visible in the provided figure.

To have some rule-of-thumb design guidelines for the waveguide height selection, let us review its impact on the waveguide performance. For the rectangular cores, as the height of the substrate reduces, the mode confinement also reduces; thus, the modes can penetrate deeper into the cladding. On the other hand, the ultrathin waveguides have less amount of losses due to the sidewall roughness and they exhibit more robust performance against fabrication imperfections [38]. This is related to

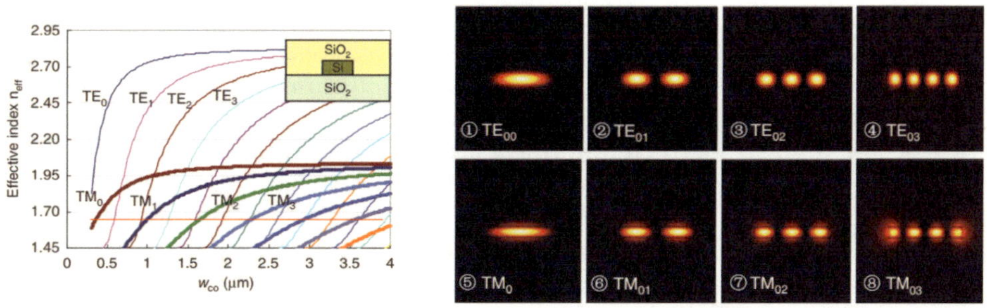

(a) (b)

Figure 2.
(a) The mode dispersion of a typical waveguide. (b) Transverse electric field distribution in different modes [36].

the fact that the roughness of the top and bottom waveguide walls can be removed by polishing; thus, the scattering losses mainly arise from sidewalls [39]. In general, thin waveguides operate at the single dominant TE mode and they exhibit good compromise in compactness (due to the lower bending radius), propagation loss, and detrimental nonlinear effect [40]. By using the 220 nm silicon on insulator (SOI) platform, the standard complementary metal-oxide-semiconductor (CMOS) process can be used for the fabrication [41]. The strain issues may be considered when choosing the layer thickness [42].

Given the waveguide aspect ratio, TE modes are more confined than TM modes and exhibit a larger effective index. Thus, in the latter case, a non-negligible part of the wave propagates in the cladding due to the delocalized nature of the TM wave fields. Note that the TE and TM waves are respectively laterally and vertically confined [43]. Birefringence, the difference of the TE and TM effective refractive indexes, is typically used to characterize the polarization sensitivity of the wave propagation in waveguides [44]. Another parameter that can aid in quantifying the localizations is the effective area [45]. Furthermore, the group index n_g of the waveguide modes can be calculated using the associated mode effective index n_{eff} as [46]:

$$n_g = n_{eff}(\lambda) - \lambda \frac{dn_{eff}}{d\lambda} \tag{1}$$

where λ is the wavelength and $dn_{eff} / d\lambda$ the first derivative of the effective index with respect to wavelength.

Rib waveguide technology (**Figure 1d**) is another promising geometry for integrated photonic devices. In this configuration, the etching depth can be used to control the scattering loss due to the side wall roughness. In this regard, rib waveguides with single and double etching steps can be utilized. By making the sidewall boundary of the bottom etch in the former case far away from the modal field, it may have little influence on modal fields [47]. Moreover, to meet the single-mode propagation condition, the height ratio of the rib waveguide might satisfy a specific condition, that is, to be larger than 0.5 in the SOI technology. The height ratio is defined as the height of the base waveguide to the total height. Also, by designing rib waveguides with large cross sections, fiber-waveguide coupling loss can be reduced [48]. The etching depth can also be used to suppress the higher-order modes by differential leakage loss [49].

As another category, subwavelength gratings (SWGs) are periodic or aperiodic geometries, where diffraction effects are suppressed by using a subwavelength grating pitch (Λ). SWGs have been employed in planar waveguides for fiber-chip surface grating couplers, microphotonic waveguides, lenses, waveguide crossings, fiber-chip edge couplers, wavelength multiplexers, ultrafast optical switches, dispersion engineering in integrated optics, and colorless directional couplers [50]. **Figure 3c** and **d** illustrates different Bragg grating waveguides with the periodic spatial modulations respectively in waveguide height, waveguide width, slab width, and cladding [51]. Bragg grating waveguides are commonly designed for TE mode operation. By using the TM modes in the design, various advantages including high quality and smoother spectral response, low propagation loss, negative group delay slope for dispersion compensation, and low coupling coefficient can be achieved [52].

Augmented low-index guiding (ALIG) waveguides consist of a low-index substrate (silica), a low-index cladding layer (air or silica), and a two-layer core made up of two dielectric materials with a high-index contrast (silicon and silicon nitride). This waveguide is based on a silicon nitride-enhanced silicon-on-insulator platform, where the TM mode is mostly confined in the low-index silicon nitride, while the TE mode is confined in the high-index silicon [53].

2.2 Waveguide design platforms

Before proceeding to bend waveguide geometries, let us review the common materials incorporated in the photonic integrated technology and discuss their advantages and disadvantages. There are several platforms for photonic integrated circuit design including silicon-on-insulator (SOI), III-V semiconductors, silicon nitride (Si_3N_4), lithium niobate ($LiNbO_3$), polymers, and rare-earth-ion (RE^{3+})-doped materials. Among all, the SOI platform, the InP platform, and the Si_3N_4 platform offer generic fabrication processes. They are commercially accessible *via* multi-project wafer (MPW) runs [54]. An ideal platform possibly combines the efficiency of electrically pumped optical sources available in GaAs and InP systems, with high-contrast Si waveguide structures or ultralow-loss waveguides available in the Si_3N_4 platform [9]. Also, because of its strong electro-optical coefficient, lithium niobate on insulator (LNOI) is another promising PIC platform to increase single-channel data transmission speed. This CMOS-compatible platform supports both monolithic and hybrid integration [55].

Silicon nitride (Si_3N_4) is gradually becoming an attractive dielectric waveguide material due to its low propagation loss, wide transparent window for wavelength from 0.4 μm up to 2.35 μm, CMOS-compatible fabrication, and large bandgap of about 5 eV resulting in negligible two-photon absorption (TPA). TriPleX ™

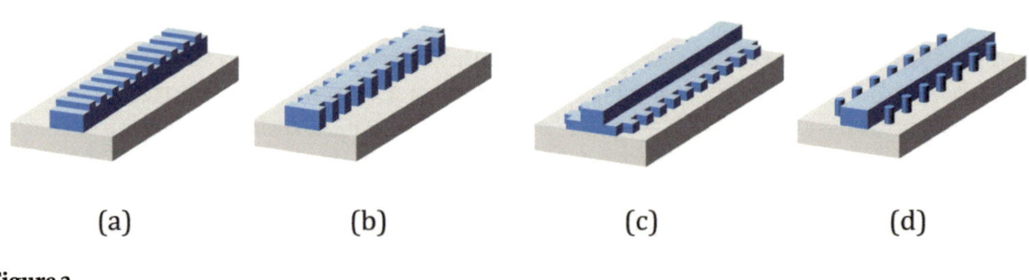

Figure 3.
Bragg grating waveguide with different spatial modulation: (a) waveguide height, (b) waveguide width, (c) slab width, and (d) cladding [51].

waveguide technology offers three main waveguides, including the box shape, and double-strip and single-strip waveguides [56]. The double-strip configuration provides a reasonable balance between the integration density and insertion loss [57]. The Si_3N_4 can be used for Kerr nonlinear photonics, and it has a large third-order nonlinear refractive index [58, 59].

The extreme polarization-dependent performance of the SOI devices demands polarization management schemes [43]. Note that there is the possibility of integrating the Si_3N_4 platform in the SOI platform or SOI waveguides in the Si_3N_4 platform to benefit from the functionalities of both [60]. Employing silicon oxynitride (SiON) core and silicon oxide cladding layers, the index contrast of the waveguide can be adjusted by varying the nitrogen-to-oxygen ratio. By this approach, the core index can be continuously varied for applications ranging from the low contrast demanded for fiber-matched processes to the high contrast required for compound semiconductor structures [61]. **Table 1** summarizes the waveguide parameters for different material categories including low/medium/high index contrast materials, $SiON_x$, III/V materials, and SOI [6].

Note that the unwanted ambient medium fluctuation may impact the performance of PICs due to the positive thermo-optic coefficient (TOC) of Si, SiO_2, and Si_3N_4, making it more critical to consider the temperature variations [62]. The effective index and TOC of the guided modes increase monolithically by increasing the waveguide width [63]. In the polymer-silica combination, the sign of the thermo-optic coefficients of the polymer and silica are opposite and the silicon substrate performs as a heat sink due to its high thermal conductivity [64].

Regarding the reconfiguration capability, the TOC of silicon nitride is much lower than silicon, which results in lower thermal sensitivity. Furthermore, the silicon nitrite-based waveguides suffer from larger dimensions due to moderate index contrast, but their phase errors are smaller for a specific fabrication error [63]. Since silicon nitrite-based reconfigurable devices have high power consumption due to their low thermo-optical coefficient, the suspended thermal phase shifters offer higher thermal efficiency and thus result in more compact devices [65].

Column	**1**	**2**	**3**	**4**	**5**	**6**
Characteristics	SiO_2 Low Δ	SiO_2 Low Medium Δ	SiO_2 High Δ	$SiON_x$	III/V	SOI
Index difference Δ (%) $\Delta = \frac{n_{core} - n_{clad}}{n_{core}}$	0.3	0.45	0.75	3.3	7.0 (46)	41 (46)
Core size (μm)	8 × 8	7 × 7	6 × 6	3 × 2	2.5 × 0.5(0.2 × 0.5)	0.2 × 0.5 0.3 × 0.3
Loss (dB/cm)	<0.01	0.02	0.04	0.1	2.5–3.5	1.8–2.0
Coupling loss (dB/point)	<0.1	0.1	0.4	3.7 (2)	5	6.8 (0.8)
Bending radius (mm)	25	15	5	0.8	0.25 (0.005)	0.002–0.005

Table 1.
Waveguide parameters for different material categories [6].

2.3 Bend waveguide design

Bend waveguides are key elements for the design of optical elements such as directional couplers, modulators, ring resonators, and interferometers. The curvature always attenuates the guided modes; thus, the geometry always suffers from radiation losses. Two main approaches to attack this problem are to either use decreasing curvatures or increase the mode confinement. These methods respectively lead to large devices and low fiber coupling efficiency [66]. Furthermore, the refractive index contrast of the core and clad impacts the amount of the bending [67]. Generally, by increasing the length of the waveguide and the refractive index contrast, the radiation losses can be reduced. Thus, in the photonic integrated circuits with compact waveguide routing, silicon on the insulator platform is preferred to silicon nitrite, since to reduce the mode mismatch loss in the bending regions in the low-loss Si_3N_4 waveguides, a large bend radius is required [68, 69]. Distortion of the transverse field profile, coupling between orthogonal polarization modes, and change of the phase constant are some other negative impacts of the curvatures on the waveguide performance, which can be controlled in carefully engineered devices.

To gain a vision of the curved structure's loss performance, a quasi-analytic technique based on the integration of a phenomenological absorption coefficient is proposed. The total loss is a function of the integral of the attenuation constant α at the local position s along the bend length S through [70]:

$$loss = \frac{10}{\log_e(10)} \int_0^S \alpha(s)ds \ (dB) \tag{2}$$

In general, the method applies to any shape function f that is continuous in its first derivative, such as sinusoidal bend [69, 70]. In this method, the bending loss per unit is represented by an exponential function [71]. The bending loss depends on the mode order in the multimode waveguides as [72]:

$$\alpha = K\exp(-cR), \text{where } c = \beta\left(\frac{2\Delta n_{eff}}{n_{eff}}\right)^{3/2} \tag{3}$$

The value of K depends on the refractive index of the core and cladding, and also the thickness of the waveguide; Δ_{neff} is the difference between the cladding refractive index and modal effective index n_{eff}, and R is the radius. Thus, as the bending radius or the mode order increases, the bending loss decreases [72].

Tapered cosine S-band may ensure better matching to the input array and connecting straight waveguides at their input ensures the single-mode propagation [30]. Moreover, by using spline bend waveguides, the losses can be shrunk with respect to circular bends, but the overall footprint remains in the same order. By introducing a small spline section at the strait-bend interface, the geometry can also be made compact. Creating the whole bend with multiple offset bends is another method for reducing the bending loss, which performs based on the reduction of the mode mismatch. The operational mechanism of this configuration is based on modulating the light path by the offset δ. Moreover, a 90° bend constructed by three sections, decorated by small offsets with each other, can be found in Ref. [73]. The problem with this technique is its performance sensitivity to the fabrication tolerance. Moreover, by optimizing the entire bend the losses can be minimized. This approach demands calculations for different bend angles and radii [74].

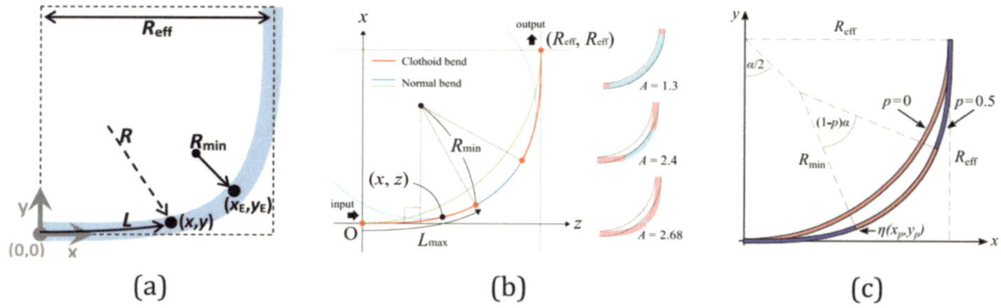

Figure 4.
Different types of Euler bends: (a) 90° normal Euler bend [75], (b) a normal bend terminated to the Euler bends in the output and input [76], and (c) partial Euler bend formed by an Euler bend terminated with normal bends [77] (reprinted with permission from aforementioned references © Optical Society of Optica Publishing Group).

The Euler bends have found applications in various integrated photonic devices. Since the eigenmode bases of straight and curved waveguides are different, the abrupt transition from infinite (in the straight section) to finite (in the bending section) curvature radius leads to spatial mode coupling. The bending-induced mode mismatch can be suppressed by the Euler bend [58]. The Euler curve, as shown in **Figure 4a**, also called the clothoid curve, has different radii of curvature in the bending path and results in low loss and low cross-talk [75]. Yet, the pure Euler curve is not an efficient way of bend waveguide design. To improve the performance, waveguide bends made up of the combinations of clothoid and normal curves are utilized for dense optical interconnect in photonic integrated circuits and they exhibit up to one order of magnitude reduction in the bending loss concerning the normal bend. The method is based on using the Euler curve on the input and output of the normal curve, where the ratio of the curves is achieved by optimization. This geometry can be found in **Figure 4b** [76]. In another method, the so-called partial Euler bend, the constant curvature waveguide input and output bands are connected to the Euler bend to balance the transition losses arising from a changing curvature and radiative losses inherent to a bend waveguide mode. The geometry of the partial Euler bend is shown in **Figure 4c** [77].

Finally, Intel, IBM, Luxtera, A*STAR, GLOBALFOUNDRIES, INPHOTEC, TowerJazz, LioniX, SMART Photonics, LIGENTEC, and Infinera are some of the companies developing photonic integrated technologies [78].

3. Conclusion

Optical waveguides are the building blocks of the photonic integrated circuits. By properly engineering the waveguide geometry, associated materials, the connecting curved waveguides, and fabrication method, it is feasible to reach low-loss devices. The designed low-loss structures may be efficiently used in conjunction with single-photon sources and detectors to conduct quantum computation. The final device potentially speeds up the calculation and gains the ability to perform calculations with some optimization strategies based on artificial intelligence.

Conflict of interest

The author declares no conflict of interest.

Author details

Shiva Hayati Raad
Independent Scientist, Tehran, Iran

*Address all correspondence to: shiva.hayati@modares.ac.ir

References

[1] Paesani S, Laing A. Quantum processors in silicon photonics. In: Silicon Photonics IV: Innovative Frontiers. Switzerland: Springer; 2021. pp. 449-489

[2] Bauters JF et al. Ultra-low-loss high-aspect-ratio Si3N4 waveguides. Optics Express. 2011;**19**(4):3163-3174

[3] Hong S, Zhang L, Wang Y, Zhang M, Xie Y, Dai D. Ultralow-loss compact silicon photonic waveguide spirals and delay lines. Photonics Research. 2022;**10**(1):1-7

[4] Thompson MG, Politi A, Matthews JC, O'Brien JL. Integrated waveguide circuits for optical quantum computing. IET Circuits, Devices & Systems. 2011;**5**(2):94-102

[5] Pavicic M. Quantum Computation and Quantum Communication: Theory and Experiments. New York: Springer Science & Business Media; 2006

[6] Thylén L, Wosinski L. Integrated photonics in the 21st century. Photonics Research. 2014;**2**(2):75-81

[7] Huang H-L, Wu D, Fan D, Zhu X. Superconducting quantum computing: A review. Science China Information Sciences. 2020;**63**:1-32

[8] Moreira RL et al. Integrated ultra-low-loss 4-bit tunable delay for broadband phased array antenna applications. IEEE Photonics Technology Letters. 2013;**25**(12):1165-1168

[9] Tran MA, Huang D, Komljenovic T, Peters J, Malik A, Bowers JE. Ultra-low-loss silicon waveguides for heterogeneously integrated silicon/III-V photonics. Applied Sciences. 2018;**8**(7):1139

[10] Lifante G. Integrated Photonics: Fundamentals. United States: John Wiley & Sons; 2003

[11] Bogaerts W et al. Programmable photonic circuits. Nature. 2020;**586**(7828):207-216

[12] De Vita C, Klitis C. Si3N4 platform for visible range applications. Riunione Nazionale di Elettromagnetismo. 2020;**2020**:1-1

[13] Wang J, Sciarrino F, Laing A, Thompson MG. Integrated photonic quantum technologies. Nature Photonics. 2020;**14**(5):273-284

[14] Elhoushi MMM. Modeling a Quantum Computer. Egypt: Ain Shams University; 2011

[15] Xie J, Zhou L, Zou Z, Wang J, Li X, Chen J. Continuously tunable reflective-type optical delay lines using microring resonators. Optics Express. 2014;**22**(1):817-823

[16] Khan S, Baghban MA, Fathpour S. Electronically tunable silicon photonic delay lines. Optics Express. 2011;**19**(12):11780-11785

[17] Zhu D et al. Integrated photonics on thin-film lithium niobate. Advances in Optics and Photonics. 2021;**13**(2):242-352

[18] Tatoli T, Conteduca D, Dell'Olio F, Ciminelli C, Armenise MN. Graphene-based fine-tunable optical delay line for optical beamforming in phased-array antennas. Applied Optics. 2016;**55**(16):4342-4349

[19] Armin F, Coia C, Nabki F, Ménard M. Compact silicon nitride interferometers. Optics Express. 2023;**31**(10):16920-16928

[20] Wu X, Fan T, Eftekhar AA, Adibi A. High-Q microresonators integrated with microheaters on a 3C-SiC-on-insulator platform. Optics Letters. 2019;**44**(20):4941-4944

[21] Xie J, Zhou L, Li Z, Wang J, Chen J. Seven-bit reconfigurable optical true time delay line based on silicon integration: Erratum. Optics Express. 2014;**22**(21):25516-25516

[22] Duan F, Chen K, Chen D, Yu Y. Low-power and high-speed 2× 2 thermo-optic MMI-MZI switch with suspended phase arms and heater-on-slab structure. Optics Letters. 2021;**46**(2):234-237

[23] Yurtsever G, Považay B, Alex A, Zabihian B, Drexler W, Baets R. Photonic integrated Mach-Zehnder interferometer with an on-chip reference arm for optical coherence tomography. Biomedical Optics Express. 2014;**5**(4):1050-1061

[24] Selvaraja SK, Sethi P. Review on optical waveguides. Emerging Waveguide Technology. 2018;**95**:458

[25] Sun X, Aitchison JS, Mojahedi M. Realization of an ultra-compact polarization beam splitter using asymmetric MMI based on silicon nitride/silicon-on-insulator platform. Optics Express. 2017;**25**(7):8296-8305

[26] Prokhodtsov A et al. Silicon nitride Mach-Zehnder interferometer for on-chip quantum random number generation. Journal of Physics: Conference Series. 2020;**1695**(1):012118

[27] Dai D, Bowers JE. Novel ultra-short and ultra-broadband polarization beam splitter based on a bent directional coupler. Optics Express. 2011;**19**(19):18614-18620

[28] Fang Q et al. Low loss fiber-to-waveguide converter with a 3-D functional taper for silicon photonics. IEEE Photonics Technology Letters. 2016;**28**(22):2533-2536

[29] Chen Z. Optical waveguide crosstalk SPICE modeling for package system signal integrity simulation. In: 2015 IEEE 65th Electronic Components and Technology Conference (ECTC). San Diego, California, USA: IEEE; 2015. pp. 2003-2011

[30] Mu J, Vázquez-Córdova SA, Sefunc MA, Yong Y-S, García-Blanco SM. A low-loss and broadband mmi-based multi/demultiplexer in Si3N4/SiO2 technology. Journal of Lightwave Technology. 2016;**34**(15):3603-3609

[31] Han L, Liang S, Zhu H, Zhang C, Wang W. A high extinction ratio polarization beam splitter with MMI couplers on InP substrate. IEEE Photonics Technology Letters. 2015;**27**(7):782-785

[32] Lecture 4: Optical waveguides. Available from: https://www-eng.lbl.gov/~shuman/NEXT/CURRENT_DESIGN/TP/FO/Lect4-Optical%20 waveguides.pdf

[33] Chen C-L. Foundations for Guided-Wave Optics. United States: John Wiley & Sons; 2006

[34] Chiang KS. Dispersion characteristics of strip dielectric waveguides. IEEE Transactions on Microwave Theory and Techniques. 1991;**39**(2):349-352

[35] Smith JA, Francis H, Navickaite G, Strain MJ. SiN foundry platform for high performance visible light integrated photonics. Optical Materials Express. 2023;**13**(2):458-468

[36] Li C, Liu D, Dai D. Multimode silicon photonics. Nano. 2019;**8**(2):227-247

[37] Gallacher K, Griffin PF, Riis E, Sorel M, Paul DJ. Silicon nitride waveguide polarization rotator and polarization beam splitter for chip-scale atomic systems. APL Photonics. 2022;**7**(4):046101

[38] Zou Z, Zhou L, Li X, Chen J. 60-nm-thick basic photonic components and Bragg gratings on the silicon-on-insulator platform. Optics Express. 2015;**23**(16):20784-20795

[39] Barwicz T, Haus HA. Three-dimensional analysis of scattering losses due to sidewall roughness in microphotonic waveguides. Journal of Lightwave Technology. 2005;**23**(9):2719

[40] Wang X et al. Continuously tunable ultra-thin silicon waveguide optical delay line. Optica. 2017;**4**(5):507-515

[41] Sia JXB et al. Compact silicon photonic hybrid ring external cavity (SHREC)/InGaSb-AlGaAsSb wavelength-tunable laser diode operating from 1881-1947 nm. Optics Express. 2020;**28**(4):5134-5146

[42] Yin M, Yang W, Li Y, Wang X, Li H. CMOS-compatible and fabrication-tolerant MMI-based polarization beam splitter. Optics Communications. 2015;**335**:48-52

[43] Saber MG, Plant DV, Abadía N. Broadband all-silicon hybrid plasmonic TM-pass polarizer using bend waveguides. AIP Advances. 2021;**11**(4):1-8

[44] Al-Hetar AM, Mohammad AB, Supa'At ASM, Shamsan ZA. MMI-MZI polymer thermo-optic switch with a high refractive index contrast. Journal of Lightwave Technology. 2010;**29**(2):171-178

[45] Lamy M et al. Si-rich Si nitride waveguides for optical transmissions and toward wavelength conversion around 2 μm. Applied Optics. 2019;**58**(19):5165-5169

[46] Zhao Z et al. Eight-channel LAN WDM (De) multiplexer based on cascaded Mach–Zehnder interferometer on SOI for 400GbE. Photonics. 2022;**9**(4):252

[47] Xiao Z et al. Design and characterization of low loss 50 picoseconds delay line on SOI platform. Optics Express. 2013;**21**(18):21285-21292

[48] Wei H, Yu J, Liu Z, Zhang X, Shi W, Fang C. Fabrication of 4 × 4 tapered MMI coupler with large cross section. IEEE Photonics Technology Letters. 2001;**13**(5):466-468

[49] Darmawan S, Lee S-Y, Lee C-W, Chin M-K. A rigorous comparative analysis of directional couplers and multimode interferometers based on ridge waveguides. IEEE Journal of Selected Topics in Quantum Electronics. 2005;**11**(2):466-475

[50] Maese-Novo A et al. Wavelength independent multimode interference coupler. Optics Express. 2013;**21**(6):7033-7040

[51] Zhou L, Wang X, Lu L, Chen J. Integrated optical delay lines: A review and perspective. Chinese Optics Letters. 2018;**16**(10):101301

[52] Chen Z et al. Spiral Bragg grating waveguides for TM mode silicon photonics. Optics Express. 2015;**23**(19):25295-25307

[53] Sun X, Alam M, Aitchison J, Mojahedi M. Compact and broadband polarization beam splitter based on a silicon nitride augmented low-index guiding structure. Optics Letters. 2016;**41**(1):163-166

[54] Mu J. Active Passive Monolithic Platforms in Si_3N_4 Technology. Netherlands: University of Twente; 2019

[55] Liu Y et al. C-band four-channel CWDM (de-) multiplexers on a thin film lithium niobate–silicon rich nitride hybrid platform. Optics Letters. 2021;**46**(19):4726-4729

[56] Yang H, Zheng P, Hu G, Zhang R, Yun B, Cui Y. A broadband, low-crosstalk and low polarization dependent silicon nitride waveguide crossing based on the multimode-interference. Optics Communications. 2019;**450**:28-33

[57] Yang H, Zheng P, Liu P, Hu G, Yun B, Cui Y. Design of polarization-insensitive 2× 2 multimode interference coupler based on double strip silicon nitride waveguides. Optics Communications. 2018;**410**:559-564

[58] Ji X et al. Compact, spatial-mode-interaction-free, ultralow-loss, nonlinear photonic integrated circuits. Communications Physics. 2022;**5**(1):84

[59] Mei C, Yuan J, Wang K, Sang X, Yu C. Chirp-free spectral compression of parabolic pulses in silicon nitride channel waveguides. In: 2016 21st OptoElectronics and Communications Conference (OECC) held jointly with 2016 International Conference on Photonics in Switching (PS). Niigata, Japan: IEEE; 2016. pp. 1-3

[60] Sacher WD et al. Polarization rotator-splitters and controllers in a Si 3 N 4-on-SOI integrated photonics platform. Optics Express. 2014;**22**(9):11167-11174

[61] Bona G-L, Germann R, Offrein BJ. SiON high-refractive-index waveguide and planar lightwave circuits. IBM Journal of Research and Development. 2003;**47**(2-3):239-249

[62] Kumari S, Gupta S. Study on temperature sensitivity of Si 3 N 4 cladded silicon 2× 2 MMI coupler. In: 2017 Fourth International Conference on Opto-Electronics and Applied Optics (Optronix). India: IEEE; 2017. pp. 1-5

[63] Tao S et al. A thermal 4-channel (de-) multiplexer in silicon nitride fabricated at low temperature. Photonics Research. 2018;**6**(7):686-691

[64] Sun X et al. A multimode interference polymer-silica hybrid waveguide 2× 2 thermo-optic switch. Optica Applicata. 2010;**40**(3):737-745

[65] Lin S, Wu Z, Zhang Y, Yu S. Compact and low-power programmable multiport interferometer on CMOS-compatible Deuterated silicon nitride. In: 2022 Asia Communications and Photonics Conference (ACP). Shenzhen, China: IEEE; 2022. pp. 1527-1529

[66] Syahriar A, Syahriar NR, Djamal JS, Saleh RR. A general S-bend approximation by cascading multiple sections of uniformly curved waveguides. Journal of Physics: Conference Series. 2019;**1196**(1):012030

[67] Heck MJ, Bauters JF, Davenport ML, Spencer DT, Bowers JE. Ultra-low loss waveguide platform and its integration with silicon photonics. Laser & Photonics Reviews. 2014;**8**(5):667-686

[68] Hai MS, Leinse A, Veenstra T, Liboiron-Ladouceur O. A thermally tunable 1× 4 channel wavelength demultiplexer designed on a low-loss Si3N4 waveguide platform. Photonics. 2015;**2**(4):1065-1080

[69] Devayani IJ, Syahriar A, Astharini D. Characteristics of S-bend optical waveguides based on back-to-back and sinusoidal structures. In: 2014 International Conference on Electrical

Engineering and Computer Science (ICEECS). Denpasar, Indonesia: IEEE; 2014. pp. 65-68

[70] Syahriar A. A simple analytical solution for loss in S-bend optical waveguide. In: 2008 IEEE International RF and Microwave Conference. Kuala Lumpur, Malaysia: IEEE; 2008. pp. 357-360

[71] Mustieles F, Ballesteros E, Baquero P. Theoretical S-bend profile for optimization of optical waveguide radiation losses. IEEE Photonics Technology Letters. 1993;**5**(5):551-553

[72] Kim D-H, Jeon S-J, Lee J-S, Hong S-H, Choi Y-W. Novel S-Bend resonator based on a multi-mode waveguide with mode discrimination for a refractive index sensor. Sensors. 2019;**19**(16):3600

[73] Ladouceur F, Labeye E. A new general approach to optical waveguide path design. Journal of Lightwave Technology. 1995;**13**(3):481-492

[74] Bogaerts W, Selvaraja SK. Compact single-mode silicon hybrid rib/strip waveguide with adiabatic bends. IEEE Photonics Journal. 2011;**3**(3):422-432

[75] Jiang X, Wu H, Dai D. Low-loss and low-crosstalk multimode waveguide bend on silicon. Optics Express. 2018;**26**(13):17680-17689

[76] Fujisawa T, Makino S, Sato T, Saitoh K. Low-loss, compact, and fabrication-tolerant Si-wire 90 waveguide bend using clothoid and normal curves for large scale photonic integrated circuits. Optics Express. 2017;**25**(8):9150-9159

[77] Vogelbacher F, Nevlacsil S, Sagmeister M, Kraft J, Unterrainer K, Hainberger R. Analysis of silicon nitride partial Euler waveguide bends. Optics Express. 2019;**27**(22):31394-31406

[78] Lelit M et al. Passive photonic integrated circuits elements fabricated on a silicon nitride platform. Materials. 2022;**15**(4):1398

Chapter 7

Impact of Process Variations on the Performance of Wavelength-Dependent Devices

Inês Venâncio, Joana Tátá, João Santos, Paulo Duarte, Carla Rodrigues and António Teixeira

Abstract

One of the main factors for the rapid growth and industrialization of the integrated photonics field was the previous development by the microelectronics sector of lithography tools and processes for complementary metal oxide semiconductor fabrication, now also employed in the fabrication of photonic integrated circuits. However, the scarce range of methods to fine-tune photonic devices during the lithographic process results in a disparity between designed and manufactured devices. For the current state of lithographic resolution, when considering wavelength-dependent devices, such as Bragg gratings, this disparity can significantly alter the design of the component and impact its performance, as the response strongly depends on its geometry. Thus, understanding the impact of lithography in a photonic component is extremely important to minimize these variations and optimize the design process. This chapter explores the impact of manufacturing process variations, studying their effect on device geometry through a case study on contra-directional couplers, toward ultimately reducing the existing gap between designers and foundries.

Keywords: lithography, photonic devices, contra-directional couplers, silicon photonics, fabrication process variations

1. Introduction

The main purpose of this chapter is to provide the reader, in particular one who is starting their endeavors in photonic integrated circuits, with the basic concepts of lithography and how this process may influence the experimental response of the designed devices by altering their geometry through a series of effects, ultimately leading to a variation of effective refractive index of the propagated mode. We hope to establish a coherent understanding of the different challenges faced in fabrication to obtain accuracy and reliability from die to die and from wafer to wafer, and assist in empowering designers by showcasing the distinctions between foundry-used terms and typical *modus operandi*. To showcase how the entirety of the process affects device's responses, especially those that are wavelength dependent, we expose the issue through experimental data of contra-directional couplers (contra-DCs).

1.1 Lithography methods

Silicon-on-insulator (SOI) platforms have emerged as a promising technology in the field of photonics. Silicon photonics, while still in its early stages of development and commercialization, offers a number of advantages for manufacturing, namely the well-established industry of producing silicon wafers for microelectronics [1]. The integration of photonic devices onto silicon chips offers the advantages of energy efficiency, affordability, and high-volume production through the use of complementary metal oxide semiconductor (CMOS) fabrication techniques imported from the microelectronics sector [1].

Photonic integrated circuits (PICs) are commonly fabricated using electron-beam lithography (EBL) or deep-ultraviolet (DUV) lithography. In both methods, the initial step consists of depositing a photoresist on top of the silicon layer to be patterned. This photoresist will act as a protective layer for the silicon during the etching process. The workflow of the two lithography methods is depicted in **Figure 1**, and it can be noted that the only difference between them lies in the photoresist patterning.

1.1.1 Electron-beam lithography

EBL is a maskless lithography technique that uses the orientation of an electron-beam canon to imprint the intended design onto the resist [2, 3]. It is a direct writing method, meaning that each point is written in succession, so the design pattering is extremely slow and dependent on the chip footprint and density. It has several advantages such as reduced feature size and improved precision. However, its main drawback is its low production rate, which makes it unsuitable for large-scale fabrication [4]. Long structures that cross the wafer are also subject to stitching errors as a result of crossing multiple writing fields (typical writing field sizes are 100, 250, 500 μm, and 1 mm [5]), which can ultimately lead to device malfunction. As such, this technique is mainly used for research purposes to fabricate novel devices.

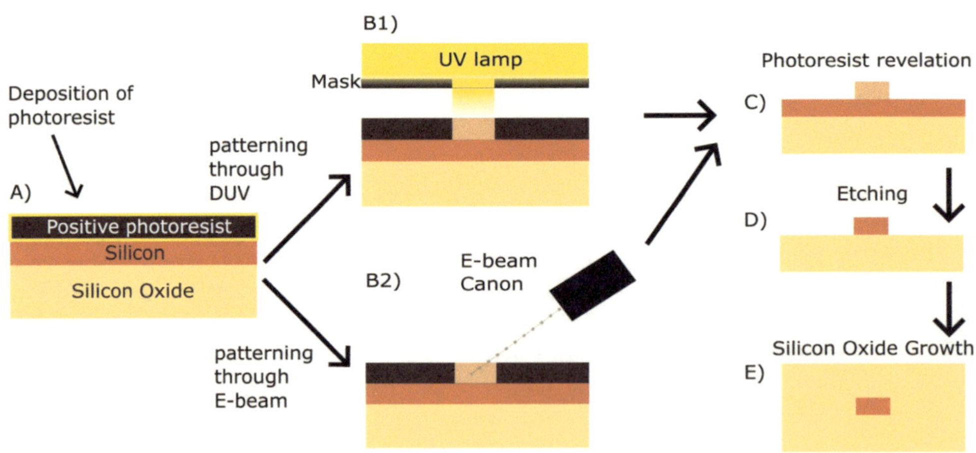

Figure 1.
Workflow of the lithography process through E-beam lithography or DUV lithography.

1.1.2 Deep-ultraviolet lithography

Contrary to EBL, DUV lithography is not a direct writing technique. It requires an intermediate step, where a chromium mask is patterned through either EBL or direct laser writing. However, once the photomask is acquired, it can be continuously used to produce chips. The photoresist layer is exposed using ultraviolet radiation, usually from UV lamps, with common wavelength values of 193 or 248 nm [6], that pass through the non-opaque features of the mask. This technique is known as projection printing because it uses a series of lenses to focus radiation onto the wafer. Compared to EBL, DUV lithography has a much bigger image field that may design an entire die in a single exposure in a few seconds. The whole wafer is patterned quickly as well with the assistance of a wafer stepper (a machine that advances the wafer to expose another portion that corresponds to a different die) [4, 7].

This fabrication technique is also the most common in CMOS manufacturing procedures. By borrowing this process, the PIC industry has the advantage of utilizing already implemented CMOS foundries, which leads to a considerable leap in the advancement of the technology compared to the primordial days of the CMOS industry [4].

The choice of which lithography process to opt for comes down to several factors. **Table 1** presents the pros and cons of the two methods explored.

1.2 Lithography errors

Despite being an extensively researched and developed technique for microelectronics and, more recently, integrated silicon photonics, DUV lithography is still prone to errors. Since the response of photonic devices is often more dependent on the structure's geometry than in the case of CMOS devices, it is critical to be aware of their potential impact.

1.2.1 Corner rounding

Corner rounding is the process of smoothing rectangular corners, which makes them less defined and more rounded. The photomask's rounded corners or feature sizes approaching or surpassing the resolution limit are the causes of this impact [8]. The existence of rectangular shapes in the original design can be significantly

Lithography method	Pros	Cons
EBL	Better resolution	Slow writing rate
	Smaller feature sizes	Unsuitable for high-volume production
	Maskless process cost-effective for low-volume applications (e.g., investigation)	
DUV	Large image field	Worse resolution
	Cost-effective for high-volume production	Larger feature sizes

Table 1.
Pros and cons of EBL and DUV lithography.

impacted by this phenomenon. Consequently, rather than the desired rectangular shape, a square profile may exhibit a sinusoidal shape [9].

1.2.2 Line shortening

Line shortening reduces the width of waveguides and may potentially cause format changes. Due mostly to diffraction, chemical species diffusion in the resist, and fluctuations in the photomask pattern, this effect becomes increasingly substantial as feature sizes approach the resolution limit [10]. This issue can have an impact on diameters and waveguide widths of a photonic device.

1.2.3 Design deviation nonlinearity

Nonlinearity refers to the difference in printed feature sizes that do not match the intended dimension. The variation imposed is usually small and insignificant for large dimensions; however, as the feature size drops, the difference between the planned and printed dimensions becomes considerably more evident. This effect is particularly noticeable for dimensions that are closer to the resolution limit. Most severely, nonlinearity may prevent some features from being printed [10].

1.2.4 Proximity effect

When feature sizes are the same as, or very close to, the lithography system's resolution limit, the proximity effect occurs, which is the variation of the dimensions printed onto the wafer as a result of environmental conditions, such as the close proximity of two waveguides. Due to their closeness, this effect causes fluctuations in the widths and distance between two close waveguides, and its significance increases as the spacing decreases [10]. This phenomenon can significantly affect the gap between two close waveguides; if the waveguides are subjected to slight fluctuations, the gap between them will subsequently change [4].

1.2.5 Sidewall roughness

In a rectangular waveguide, sidewall fluctuations arise from sidewall roughness. These fluctuations lead to slight changes in the waveguide's width, which in turn affects the waveguide's effective refractive index throughout its length [11]. This phenomenon may result in considerable phase error and propagation losses [12].

1.3 Foundry bias vs. fabrication variability

Two main causes of variability are usually present in the fabricated devices. The first one refers to the bias that is introduced by the specific foundry workflow. Foundries provide a *Design Manual* (a document containing the rules that designers must follow, and stack and layer information) with an interval of values in which this error is inserted. This type of variability is constant within the foundry runs for the same design and can be accounted for after the first iteration. Its source is the lack of in-line testing methods that enable the accurate measurement of the structures being designed, which, compared to CMOS devices, have a higher range of distinct geometric features, and cannot be easily tested in terms of functionality.

The other cause of variability lies in the errors discussed in Subsection 1.2 and the overall resolution and accuracy of the process that, all combined, contribute to the geometric and functional variance across the wafer of the designed structures. Unlike the foundry bias that can be predicted and accounted for in the design from wafer to wafer, fabrication variability is often "random" within the same wafer. Its values are low, in the order of few nanometers, and, for most structures, do not impact the functionality of the device. However, for structures that have features below the limits of resolution and are, therefore, more vulnerable to lithography errors, the geometric variability between dies (individualized sections of a wafer, repeated throughout it) increases. If that same device is wavelength dependent, with high contrast between the cladding and core refractive index, like in the case of SOI platforms, then this variance has a higher impact on the functionality of the device. That same impact will be presented and studied in Section 2.

1.4 Contra-directional couplers

A contra-DC is a device composed of two parallel waveguides in close proximity, each with periodic sidewall corrugations, combining the characteristics of a Bragg grating and a directional coupler forming a four-port device: *in*, *through*, *drop*, and *add* (**Figure 2a**).

Due to the corrugations in the waveguides, the optical signal traveling the device will experience partial reflections at every corrugation. As such, constructive interference will arise for wavelengths that satisfy the Bragg condition [13], as shown in **Figure 2b**. In the case of the contra-DC, this condition is expressed by:

$$\beta_1 + \beta_2 = \frac{2\pi}{\Lambda} \tag{1}$$

where Λ is the pitch of the corrugations and $\beta_{1,2}$ are the propagation constants of the waveguides given by:

$$\beta_{1,2} = \frac{2\pi n_{\text{eff}1,2}}{\lambda} \tag{2}$$

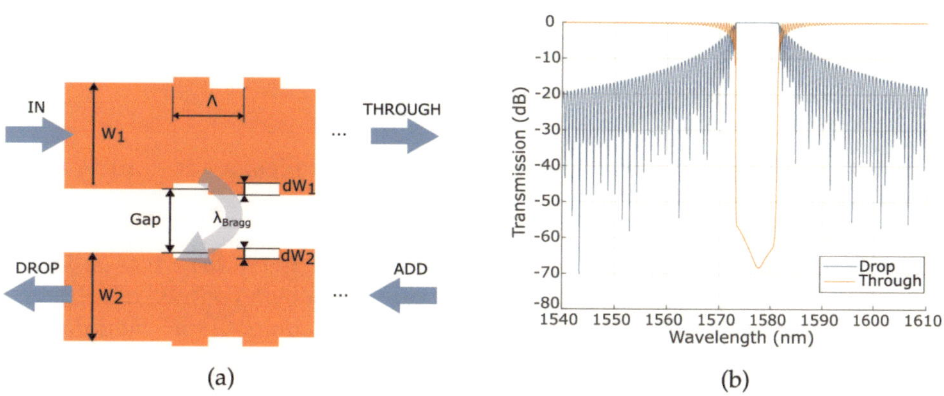

Figure 2.
(a) Structure diagram and (b) typical simulated spectral response of a contra-DC.

where $n_{\text{eff}1,2}$ are the effective refractive indices of the waveguides and λ is the wavelength of the propagated mode.

Contrary to a regular directional coupler designed specifically for power coupling, the waveguides of a contra-DC have different widths and, therefore, different propagation constants (expressed in Eq. (1)), thus reducing the forward-propagating co-directional power coupling between waveguides [1]. Because of this feature, only the backward-propagating mode (the signal reflected at the Bragg gratings) will couple into the second waveguide. Therefore, when a signal enters the device through the *in* port, the portion of the signal that satisfies the Bragg condition will thus exit at the *drop* port while the rest of the signal will exit at the *through* port.

Solving Eq. (1), we arrive at a simplified condition for the reflected wavelength in the case of contra-DCs:

$$\lambda_{\text{Bragg}} = (n_{\text{eff}1} + n_{\text{eff}2})\Lambda \tag{3}$$

where λ_{Bragg} is the Bragg wavelength, and $n_{\text{eff}1}$ and $n_{\text{eff}2}$ are the effective refractive indices of the first and second waveguides of the contra-DC, respectively.

As stated in Eq. (3), the central wavelength of the filtered response will depend only on the effective refractive indices of both waveguides and the pitch of the corrugations. However, for each individual waveguide, the Bragg condition for each grating can be satisfied, thus giving rise to intra-waveguide reflections. Analogously to Eq. (3), these reflected signals will have their central wavelength at $\lambda_1 = 2n_{\text{eff}1}\Lambda$ and $\lambda_2 = 2n_{\text{eff}2}\Lambda$, for the first and second waveguides, respectively. Because of this phenomenon, contra-DCs possess an anti-reflection design (already demonstrated in **Figure 2a**) that consists of the misalignment of the sidewalls that compose the Bragg gratings, phase-shifting them by 180° [1]. By misaligning the sidewalls, the signal now reflects with a π shift from sidewall to sidewall, thus making the partial reflections interfere destructively. As a result, the signals at λ_1 and λ_2 are no longer reflected to the drop port, ultimately removing undesired peaks from the spectral response of the contra-DC [14].

This device has been extensively studied by the *SIEPIC* group from the University of British Columbia in Canada, led by Professor Lukas Chrostowski [1, 15, 16]. They have put forward multiple tools to assist in photonic integrated circuit design available to the public, including a repository containing scripts for simulating a contra-DC using *Ansys Lumerical* [17], present in Appendix A, by Mustafa Hammood. Additionally to their contributions, we also leave a code of a parameterizable contra-DC in *GDS Factory* [18] for implementation of this building block in future designs in Appendix B.

2. Impact of geometry variations on photonic devices: a case study on contra-DCs

In photonics, SOI platforms are most commonly used, given the advantageous property of high-confinement waveguides [1], due to the high contrast of refractive indices of Si that compose the waveguides and SiO_2 that compose the cladding. However, this high contrast also translates into large variations of effective refractive index when the geometry of the waveguide alters.

To evaluate the effect in the optical properties of integrated waveguides on photonic structures that their geometry imposes, a study on *Ansys Lumerical MODE* solver

[17] was performed on an SOI platform with a sidewall angle of 85, considered to approximate the design to the as-fabricated waveguide.

Figure 3 shows the variation of effective refractive index (n_{eff}) of a waveguide with the variation of its width, for a thickness of 220 nm.

The impact that varying the width of a waveguide has on its n_{eff} is clearly visible. With an approximately linear variation of effective index, when one considers a Bragg device along with its phase condition, it becomes clear that if the waveguide width deviates from the designed width, so will the Bragg wavelength of the grating, which may greatly impact the performance of the device.

Figure 4 is an image taken with scanning electron microscopy (SEM) of a section of a contra-DC that was fabricated through DUV lithography. It is possible to observe a significant rounding of the corrugations whose shape becomes sinusoidal with the lithography effect. This over-etch will be responsible for a significant decrease in the bandwidth due to the increase of the gap between waveguides and the erosion of the corrugation width. It will also be responsible for a shift in the central wavelength of the contra-DCs, by the phenomenon just analyzed in **Figure 3**. Since the corrugations are created by a shift in the position of each half-period section instead of by variations of widths as part of the anti-reflection design, the effect of lithography onto the waveguide width of the corrugations section can be approximately correlated to the shift in central wavelength.

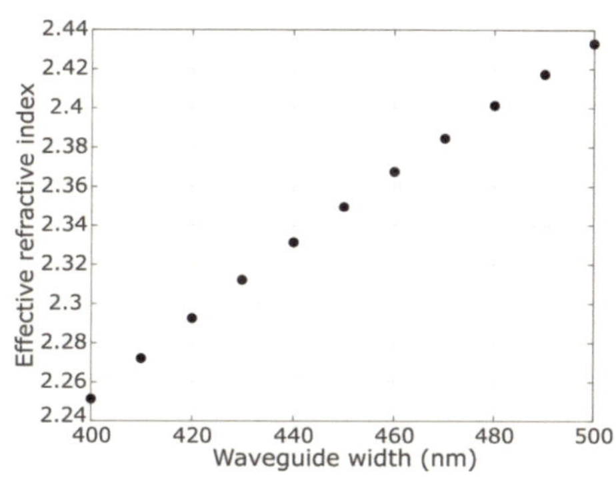

Figure 3.
Variation of effective refractive index of a Si waveguide as a function of its width.

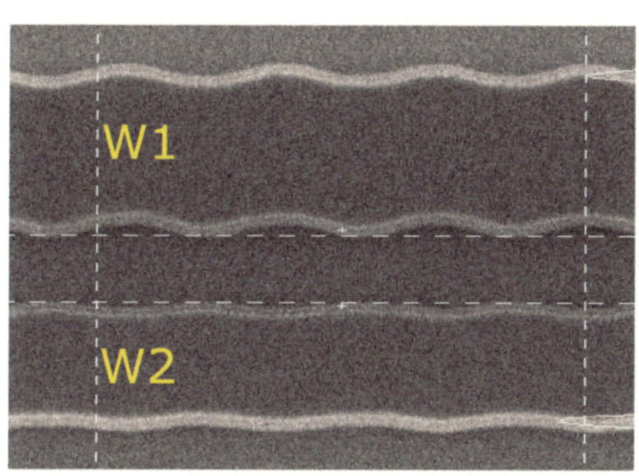

Figure 4.
SEM image of a contra-DC section.

With these results in mind, three wafers with contra-DCs fabricated through DUV lithography were characterized, and a model that relates the geometry of the waveguides with the measured optical properties of the contra-DCs was conceived and explored, which will be the focus of the next section.

2.1 Die-to-die study

To study the variation and impact of the lithography process on wavelength-dependent devices, a SOI wafer containing 72 dies with contra-DCs, fabricated through 193 nm DUV lithography, was fully characterized. The optical spectra at the drop and through ports were acquired and analyzed.

The central wavelength of the drop response of each contra-DC was mapped according to its position on the wafer. **Figure 5a** shows the central wavelength map of the wafer, where the scale is shown as the difference between the expected central wavelength from the design ($\lambda_{designed}$), which was obtained through simulation beforehand, and the measured central wavelength of the devices ($\lambda_{measured}$).

From **Figure 5a**, we can see the clear variation of central wavelength present across the wafer. The maximum redshift obtained was 26.1 nm while the maximum blueshift was −9.5 nm. Furthermore, the average difference between expected and measured central wavelength is 0.05 nm while the standard deviation is 7.7 nm.

Despite the seemingly small difference between average measured central wavelength and expected central wavelength, observing **Figure 5a**, it is visible that the redshift of the outer ring dies (contra-DCs at the edge of the wafer) is what is causing this effect while the center of the wafer seems to be predominantly blueshifted. Furthermore, it is noticeable that the outer ring of the wafer follows a much more random variation than the ones located toward the center of the wafer. This phenomenon is associated with a particular step of the fabrication process; the photoresist that is deposited on top of the Silicon layer is done so through spin coating. This technique consists of dispensing the fluid in the center of the wafer while it spins at several thousand rotations per minute. The centrifugal force associated with the rotation movement creates a laminar flow of the fluid toward the edge of the wafer, creating a uniform thin film. However, due to the surface tension at the borders, the fluid tends

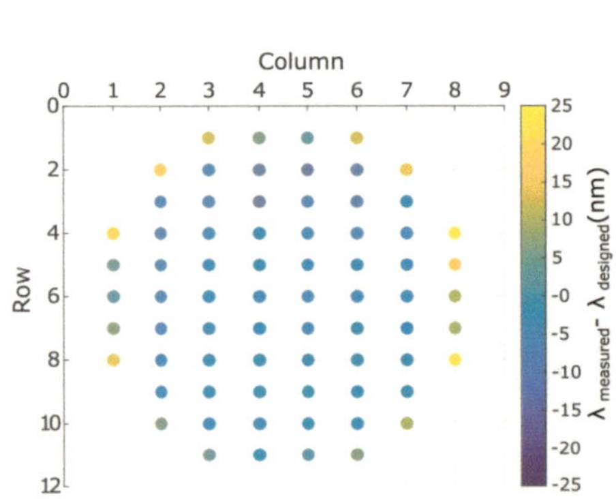

(a)

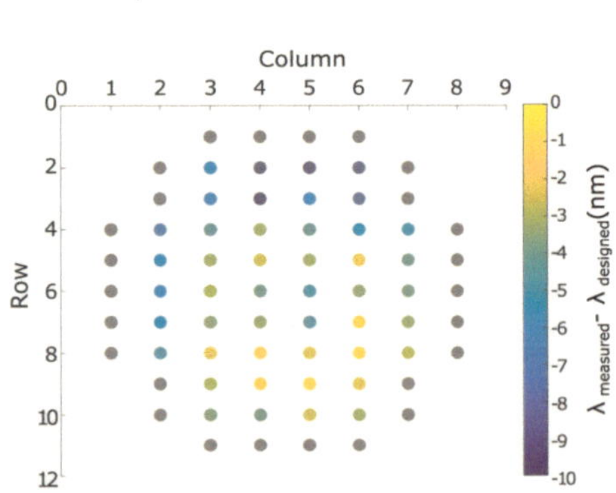

(b)

Figure 5.
Central wavelength map of (a) full wafer and (b) wafer without outer ring PICs (represented in gray).

to accumulate at the edges of the wafer, thus making the thickness of the layer in these regions usually greater [19].

Due to the thicker layer of photoresist at the edges of the wafer, when it is exposed to UV radiation, the photonic design is not etched as well as in the center region of the wafer. Because of this phenomenon, devices fabricated at the edge of wafers are commonly discarded [19].

Taking this account, the wafer without the outer ring was analyzed; the central wavelength map of the wafer is shown in **Figure 5b**, where the variation is notably less significant. Without the edge dies (represented in the figure in gray), the average difference of central wavelength becomes −4.2 nm with a significantly smaller standard deviation of 2.0 nm. Comparing these results to the full wafer analysis, we can conclude that the tendency is for the devices to be blueshifted from what was designed.

Furthermore, it is possible to determine that, contrary to the edge of the wafer, the remaining dies follow a more consistent shift from the expected value, which leads to the conclusion that the majority of the inner dies suffer mostly from foundry bias.

It should be noted that the devices characterized in this work were not the result of the first run with the foundry, and, as such, the designs already possessed a compensation factor that was induced from previous runs done with the same foundry. For this reason, the devices present a smaller central wavelength shift than usual first-run devices.

2.2 Wafer-to-wafer study

Following the die-to-die analysis presented previously, it was important to assess the variability of the performance of the devices across multiple wafers, in order to better understand the origin of the wavelength shift of the contra-DCs (as explored in Section 1.3) and evaluate the reproducibility of the results.

Dies from two additional identical wafers were characterized and analyzed, with 19 samples each. It is important to note that only dies from the inner region of the wafer were included in this analysis as a consequence of the study and conclusions presented in Section 2.1.

Figure 6 shows the results for all three wafers regarding the central wavelength of the contra-DCs, as well as the average of the three wafers for each die. In **Table 2**, the average and standard deviation of each wafer are displayed. Once again, all results are presented relative to the expected results ($\lambda_{measured} - \lambda_{designed}$).

From **Table 2**, it is clear that the shift observed in the contra-DCs is similar across all three wafers tested, with the average central wavelength of the wafers varying within less than 1 nm. Consequently, the hypothesis that the main impact is the foundry bias gains more strength. This translates into a constant difference in waveguide width from designed to fabricated, which will be explored further.

Analyzing the overall result, which is the average of the three wafers that were characterized, the central wavelength is, on average, blueshifted by 4.9 nm from the expected; this implies that, on future runs, if this figure of merit is compensated, the devices will expectedly be closer to the desired performance. Taking the foundry bias into account, the central wavelength shift can be related to the waveguide width of the devices, which can be extracted, accounted for, and corrected in future designs, provided they are manufactured in the same foundry. This study will be the focus of the next section.

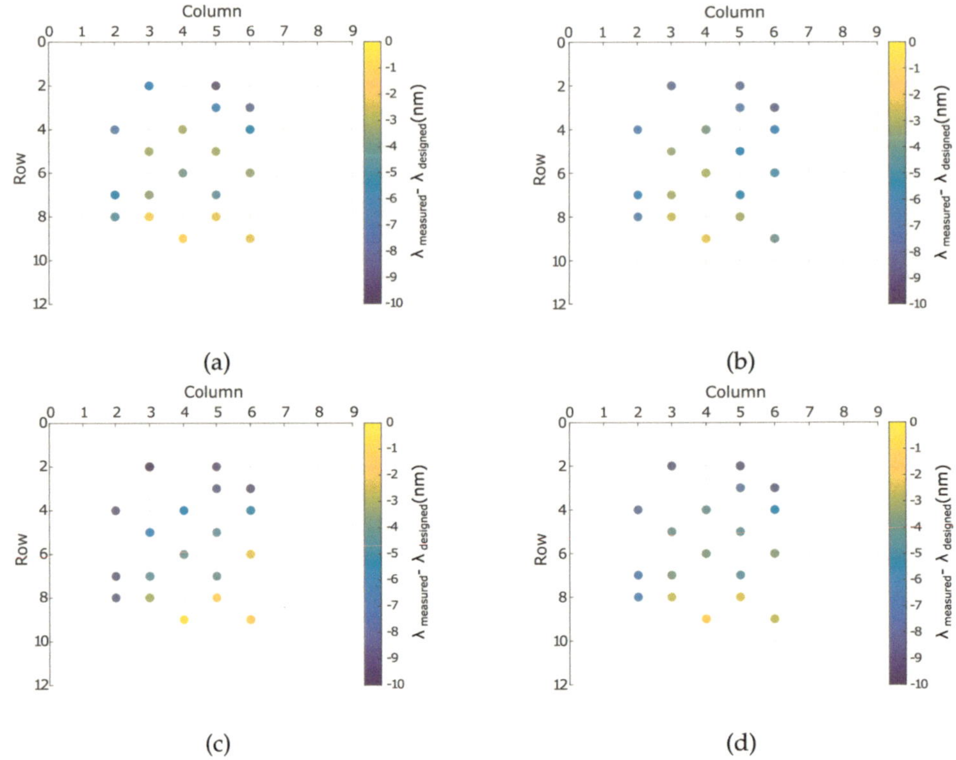

Figure 6.
Central wavelength map of (a) wafer A, (b) wafer B, (c) wafer C, and (d) wafers' average (the points present in the graphs of this figure are solely the common dies tested among all wafers to assure accuracy in the statistical analysis).

Wafer	Average	Standard deviation
	$\lambda_{measured} - \lambda_{designed}$ (nm)	$\lambda_{measured} - \lambda_{designed}$ (nm)
A	−4.4	1.9
B	−5.0	1.6
C	−5.3	2.6
Overall	−4.9	1.9

Table 2.
Central wavelength performance for all wafers.

2.3 Geometry extraction model

In integrated wavelength-dependent devices, such as contra-DCs, the source of performance deviations is geometry variations induced by the fabrication process. Therefore, it is of the utmost importance to relate these two elements in a way to optimize future designs.

Based on the laboratory results presented in Sections 2.1 and 2.2, a model relating the central wavelength of the optical response measured from a contra-DC and the waveguide width was developed. For the following analysis, the impact of the variation in thickness in the device's response was not accounted for.

Recalling Eq. (3), the central wavelength of a contra-DC can be determined by plotting the respective dispersion curve of the Fundamental transverse electric (TE) mode (where $n_{eff} = \frac{n_{eff1}+n_{eff2}}{2}$) and the Bragg condition ($\frac{\lambda}{2\Lambda}$). The intersection point between the two curves will be the theoretical central wavelength of the device.

Dispersion curves for contra-DCs with different waveguide widths were obtained for widths of $w_{1,2} = 400, 300$ nm to $w_{1,2} = 500, 400$ nm. In this study, the waveguide width of the second waveguide was always considered to be 100 nm narrower than the first waveguide (e.g., $w_1 = 500$ nm, $w_2 = 400$ nm).

The results obtained are shown in **Figure 7a**. It is clear that the effective refractive index of the device increases as the waveguides widen due to the higher influence of the Si refractive index in the medium. **Figure 7b** presents the intersection points for each dispersion curve, which directly relates to the corresponding waveguide width. Therefore, from a measured optical response, one can directly extract the waveguide widths of the device through this curve.

Using the correlation established and represented in **Figure 7b**, the waveguide width of each device on wafer A (from Section 2.1) was extracted. The outer ring of the wafer was once again discarded. The results can be observed in **Figure 8** and are presented as the difference between extracted width and designed width ($w_{extracted} - w_{designed}$).

The variation of waveguide width follows the same pattern across the wafer as the central wavelength, as expected, since the waveguide width has a nearly linear

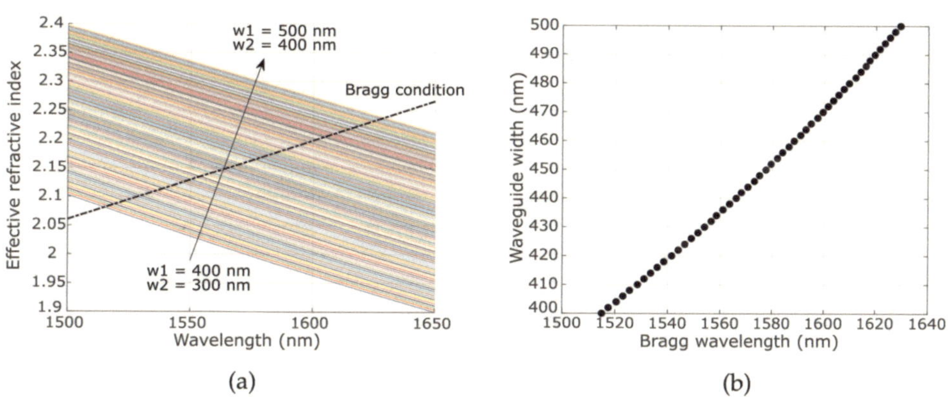

Figure 7.
(a) Dispersion curves for different waveguide widths and effect on Bragg condition match and (b) waveguide width and Bragg wavelength relation.

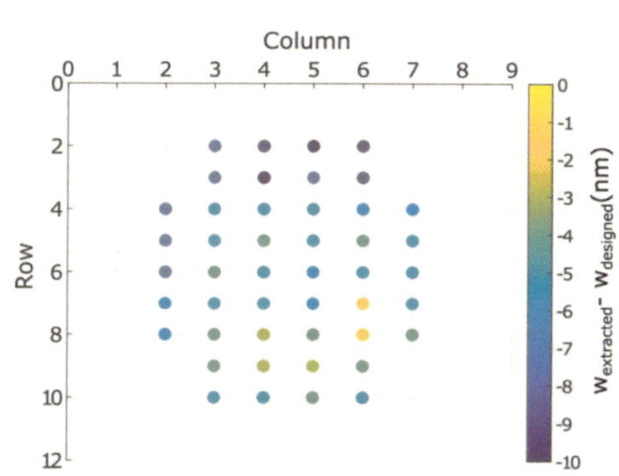

Figure 8.
Waveguide width map of wafer A.

dependency on the central wavelength (**Figure 7b**). All devices present smaller waveguide widths than the designed, with the maximum shift being −10 nm, while the average $w_{extracted} - w_{designed}$ is of −5.7 nm.

The waveguide width map was also extracted for the other wafers tested to establish a comparison between them. **Figure 9** presents the results obtained for the three wafers as well as the overall average among wafers. Once again, it is observed that all devices from all wafers present a smaller waveguide width than designed. Furthermore, the three wafers seem to present the same map pattern (small deviation in die comparison between wafers), which indicates repeatability across wafers.

As such, for future designs, two approaches can be considered: for a full wafer (private run) production, the designer may take the zone of the wafer where the device will be located into consideration for a more precise waveguide width compensation by using the precise value of the die average. Alternatively, considering a multi-project wafer (MPW) approach where the designer does not have access to the wafer region of its PIC, compensating the designs with the average of the wafer, thus increasing 6 nm in the waveguide width, would be the solution.

Despite these promising results that would minimize the wavelength shifts observed in future runs, one can also clearly see the variation from die-to-die and wafer-to-wafer, which cannot be attributed solely to foundry bias. These variations are related to the fabrication problems explored in Section 1.2. Contrary to the foundry bias-induced shift, these deviations are much harder to compensate given

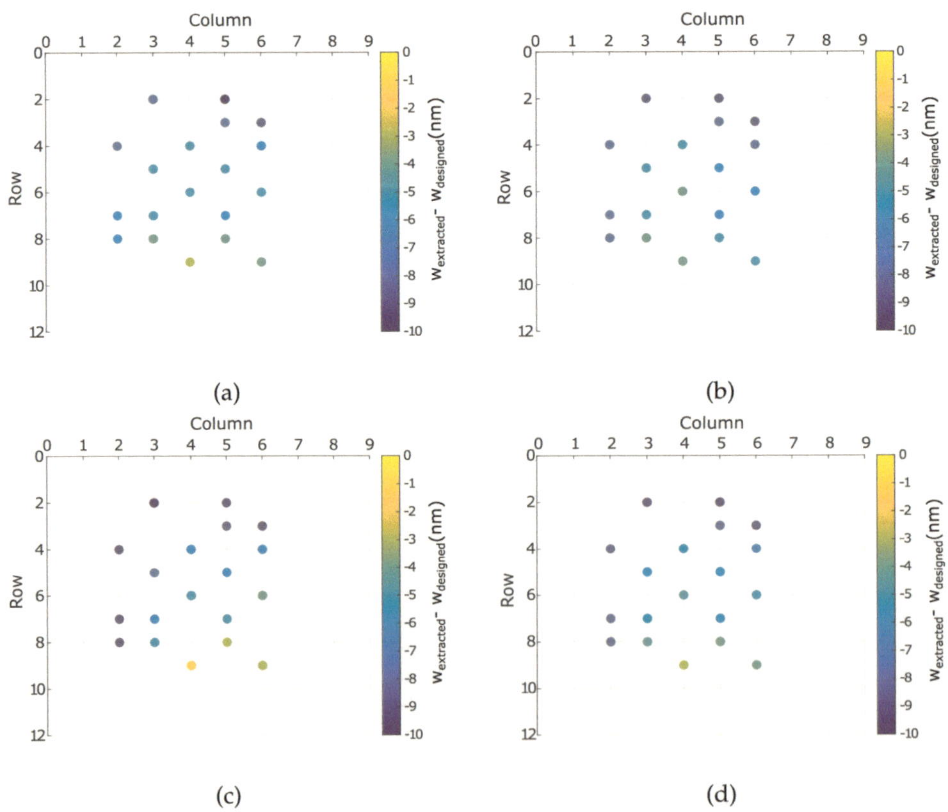

Figure 9.
Waveguide width map of (a) wafer A, (b) wafer B, (c) wafer C, and (d) wafers' average.

Wafer	Average	Standard deviation
	$w_{extracted} - w_{designed}$ (nm)	$w_{extracted} - w_{designed}$ (nm)
A	−5.7	1.6
B	−6.0	1.4
C	−6.4	2.5
Overall	−6.0	1.7

Table 3.
Extracted waveguide width for all wafers.

their randomness. Consequently, there is a limit to what the designer can do to eliminate this source of variability, and can therefore mostly take mitigation action at best, which will be the topic of the next section (**Table 3**).

3. Mitigation strategies

3.1 Fabrication

It is clear that the improvement of the lithography techniques leading to an increment of resolution would undoubtedly help to mitigate the fabrication errors that most PIC designers struggle with. Indeed, there have been advancements in lithography technology, and high-numerical aperture (NA) extreme ultraviolet (EUV) lithography technology with resolutions up to 13 nm is now a potential reality in the CMOS industry [20]. As the progress in lithography technology advances, the integrated photonics industry will also benefit from it.

As for the individual process of each foundry, in-line testing needs to become a reality to increase the reliability and reproducibility of the devices [21]. Morphological testing and critical dimension (CD) alignment are great improvement steps toward eliminating foundry bias. However, morphological testing requires the stopping of the process, which can significantly increase the fabrication time, while the CD alignment is only a viable solution for wafers with a similar layout among dies in terms of density.

Besides morphological testing, vertical grating couplers make functional testing at a wafer level possible, similar to a multi-probe station. But for novel structures, there will be a gap between the designer and fabrication engineer regarding the response of the device, especially if it is influenced by multiple parameters. The increase in the amount of structures present in each foundry's process design kit (PDK), as a result of intensive testing and fabrication correction, will ultimately allow the increase of the accuracy of the design's response, while simultaneously increasing the designer's possibilities.

3.2 Lithography simulators

Meanwhile, designers and foundries can and should collaborate to create and improve lithography simulators, which are simulation programs that allow the emulation of the effect that lithography will have on the features designed. This type of tool is crucial for developing devices that contain features with dimensions below the resolution's limits. Currently, multiple simulation companies such as Synopsys and Siemens have their own solutions (Synopsys S-Litho and Siemens Calibre) that were developed for the IC industry but that the PIC industry can also benefit from. Apart from commercial simulators, there are several open-source solutions, such as *PreFab Photonics* [22] and *ELIAS* from the University of Texas at Austin [23]. Open source or not, the main setback present in these programs is their requirements of either lithography model parameters, such as *NA* (related to the ability of the lens to collect light; range of the angular acceptance of the lens), σ (partial coherence factor; a measure of the angular range of the illumination relative to the angular acceptance of the lens) and $\delta\sigma$ (angular variation ranges from which the intensity of the optical beam rises from 0.5% to 99.5%) [4, 7] which would require in-depth knowledge of the foundry process, and therefore are not accessible to fabless designers, or of a sample of SEM images of test structures from the foundry and process of interest, typically not supplied in MPWs [4, 22].

3.2.1 PreFab photonics emulation of a contra-DC

We used *PreFab Photonics* [22] as an experimental lithography simulator to observe the prediction of the impact of fabrication on a contra-DC. The only models available at the time were from *Applied Nanotools Inc.*, which exclusively fabricates with EBL; therefore, we are not able to draw extensive conclusions, given that our devices were fabricated through DUV lithography. As expected, the corrugations of the device are the most affected, given that they fall below the resolution limit. But as seen in **Figure 10**, the corner rounding effects in the corrugations are even more severe, leading to a sinusoidal shape of the waveguide edges. This will clearly impact the central wavelength, given that the width will decrease due to the over-etch of the corrugations. The period has remained identical, so it is possible to consider the waveguide width as the main factor of the central wavelength variation.

3.3 Design as a lithography error mitigation strategy

Ultimately, all the strategies presented above are limited in mitigating the lithography variability. For devices where the variation in the structure dimensions will almost equally impact the response's central wavelength, such as in the case of the contra-DCs, designing needs to be included as a strategy to mitigate error. The clearest answer would be to limit the use of structures that do not push or are as not as affected by the resolution limits. However, for the cases where that is not possible, we present two solutions.

3.3.1 Device tuning

Tuning refers to the method of slightly varying the response of a device by modeling the refractive index of the material (clearly leading to a variation in the effective

Figure 10.
(a) Input design in the PreFab Photonics lithography emulator and (b) lithography simulation of the input design using the ANT NanoSOI v3 model.

refractive index of the propagated mode) through the change or exposure of a physical property most often represented by a constant, such as temperature (thermo-optic coefficient) or an electric field (Pockels coefficient). Regarding contra-DCs, it has been reported that 80 K would be necessary to reach a variation of central wavelength in SOI slightly higher than 5 nm [24], which renders the strategy inefficient, given the high-power consumption that the PIC would require. Wei Shi, et al. reported a 7 nm shift in wavelength with 18 V applied to a Contra-DC designed in a rib platform with a p-i-n structure for high-speed tuning or switching through carrier injection [25].

3.3.2 Designing for redundancy

Designing for redundancy means multiplying the critical components present in our circuit to enhance reliability and availability. By having several contra-DCs in place with slight parameter variations, the probability that one of them will give the pretended response is higher than to just employ a single device. This strategy would require the employment of photonic switches, allowing the system to tolerate the failure of one structure without compromising its overall functionality. However, this approach will lead to a chip with a bigger footprint and increased complexity.

4. Conclusions

In this chapter, the impact of the lithographic process on PICs and, most importantly, in wavelength-dependent devices was explored. A case study on contra-DCs was developed, and the performance of the devices, according to their position in the wafer, was studied and analyzed.

Through this case study, we found that devices located on the outer ring (at the edge) of the wafer present a significant central wavelength deviation when compared to identical devices that were fabricated toward the center of the wafer. These findings lead to the conclusion that devices manufactured in these regions of the wafer are not viable, as they tend to present a random and large performance deviation.

Excluding the contra-DCs from the wafer's edges, we observed a systematic deviation among devices when comparing them die to die and wafer to wafer, with an average of −4.9 nm between measured and expected central wavelength and a standard deviation of 1.9 nm. The results indicate the predominance of a foundry bias in the fabrication process of the characterized wafers.

In view of the characterization results obtained, a geometry extraction model was created in order to directly relate the optical response of the contra-DCs and their waveguide widths. Through this model, it was possible to induce the value of the foundry bias of the wafers in study, where the extracted widths of the devices are, on average, 6.0 nm narrower than the designed waveguides.

Regarding the variability present in the dies, foundry bias was found to be predominant. This type of variability can be decreased further by adjusting the designs according to the experimental values of each run. However, random variation is still present, as can be concluded by the value of standard deviation. Although it contributes with a much smaller value to the variability of the devices, since it is tied to the limits of resolution and present lithography errors, it will always remain an unpredictable hindrance to designers.

Overall, the study allows for a sense of the existing gap between designers and foundries. Although developments have been made regarding lithography fabrication tools that further push the limits of resolutions, with techniques such as EUV lithography, those are still hard to access. Therefore, designers must focus on solutions that allow the control or contour of the fabrication effects, such as the implementation of tuning or designing for redundancy. Increasing foundry testing of structures through morphological or functional verification should lead to optical responses of higher accuracy. The challenge is implementing them with no further significant strain during the fabrication process.

Acknowledgements

This work is supported by the European Regional Development Fund (FEDER), through the Competitiveness and Internationalization Operational Programme (COMPETE 2020) of the Portugal 2020 framework [Project POWER with Nr. 070365 (POCI-01-0247-FEDER-070365)].

Nomenclature

CD	critical dimension
CMOS	complementary metal oxide semiconductor
contra-DC	contra-directional coupler
DUV	deep-ultraviolet
NA	numerical aperture
EBL	electron-beam lithography
EUV	extreme ultraviolet
IC	integrated circuit
MPW	multi-project wafer
PDK	process design kit
PIC	photonic integrated circuit
SEM	scanning electron microscopy
SOI	silicon-on-insulator
TE	transverse electric

Appendix A

SIEPIC contra-DC simulator repository by Mustafa Hammood: https://github.com/SiEPIC/contraDC.

Appendix B

```
#    Contra-DC design in GDS factory
#    Authors: Joao Santos and Joana Tata   ; joana.tata@picadvanced.com
#    PICadvanced S.A., Portugal
#    (c)2024

import math
from functools import partial
import gdsfactory as gf
```

```
# rounding functions
def round_up(number, decimals=0):
    multiplier = 10 ** decimals
    return math.ceil(number * multiplier) / multiplier

def round_down(number, decimals=0):
    multiplier = 10 ** decimals
    return math.floor(number * multiplier) / multiplier

# ContraDC cell
@gf.cell
def simple_contra_dc(
        w1,  # width of waveguide 1
        w2,  # width of waveguide 2
        gap,  # gap between waveguides and between the middle of the
    corrugations from waveguide 1 and 2
        dw1,  # corrugation width of waveguide 1
        dw2,  # corrugation width of waveguide 2
        wg_io,  # system waveguide
        n,  # number of corrugations
        p,  # period/pitch of corrugations
        cross_section,  # core cross_section
        cross_section_port  # port cross_section
) -> gf.Component:
    sbend_h = 3  # Height of the SBend
    sbend_l = 12  # Length of the SBend
    tapper_l = 15  # Length of the taper

    # rounded length of corrugations
    length_a = round_up((p / 2), decimals=3)
    length_b = round_down((p / 2), decimals=3)
    # rounded width of corrugations
    dw1 = (round((dw1 / 4) * 1000) / 1000) * 4
    dw2 = (round((dw2 / 4) * 1000) / 1000) * 4
    # shift of the sections
    profile1 = dw1 / 2
    profile2 = dw2 / 2

    c = gf.Component("contra_dc")
    halfp_ref = []
    wgt = gf.Component(name="TopWG")
    wgb = gf.Component(name="BotWG")

    # --- Draw bragg waveguides ---
    # - draw corrugations of the Top waveguide
    for i in range(0, n, 1):
        halfp_l = gf.components.straight(
            length_a, width=w1, cross_section=cross_section, add_pins=
    False)
        halfp_ref.append(wgt << halfp_l)
        if i >= 1:
            halfp_ref[-1].xmin = halfp_ref[-2].xmax
```

```
    halfp_ref[-1].movey(profile1)
    halfp_r = gf.components.straight(
        length_b, width=w1, cross_section=cross_section, add_pins=
False)
    halfp_ref.append(wgt << halfp_r)
    halfp_ref[-1].xmin = halfp_ref[-2].xmax
    halfp_ref[-1].movey(-profile1)
    del halfp_r, halfp_l

# --- Fanout waveguides ---
# -  draw inputs and outputs of the Top waveguide
aux_cs = partial(gf.cross_section.cross_section, width=w1,
cross_section=cross_section)
bndwg = gf.components.bend_s((sbend_l, sbend_h), cross_section=aux_cs,
 add_pins=False)
inwgb_r_w1 = wgt << bndwg.mirror((0, 1), (0, 0))
aux_cs = partial(gf.cross_section.cross_section, width=w1,
cross_section=cross_section)
obndwg = gf.components.bend_s((sbend_l, sbend_h), cross_section=aux_cs
, add_pins=False)
outwgb_r_w1 = wgt << obndwg
inwgb_r_w1.connect("o1", halfp_ref[0].ports["o1"])
outwgb_r_w1.connect("o1", halfp_ref[-1].ports["o2"])

taperwgi = gf.components.taper(
    length=tapper_l,
    width2=w1,
    width1=wg_io,
    cross_section=aux_cs,
    add_pins=False)
taperwgo = gf.components.taper(
    length=tapper_l,
    width2=w1,
    width1=wg_io,
    cross_section=aux_cs,
    add_pins=False)
itappert_r = wgt << taperwgi
otappert_r = wgt << taperwgo

# - draw corrugations of the Bottom waveguide
halfp_ref = []
for i in range(0, n, 1):
    halfp_l = gf.components.straight(
        length_a, width=w2, cross_section=cross_section, add_pins=
False)
    halfp_ref.append(wgb << halfp_l)
    if i >= 1:
        halfp_ref[-1].xmin = halfp_ref[-2].xmax
    halfp_ref[-1].movey(-profile2 - gap - w1 / 2 - w2 / 2)
    halfp_r = gf.components.straight(
        length_b, width=w2, cross_section=cross_section, add_pins=
False)
```

```
    halfp_ref.append(wgb << halfp_r)
    halfp_ref[-1].xmin = halfp_ref[-2].xmax
    halfp_ref[-1].movey(profile2 - gap - w1 / 2 - w2 / 2)
    del halfp_r, halfp_l

# --- Fanout waveguides ---
# - draw inputs and outputs of the Bottom waveguide
aux_cs = partial(gf.cross_section.cross_section, width=w2,
cross_section=cross_section)
bndwg = gf.components.bend_s((sbend_l, sbend_h), cross_section=aux_cs,
 add_pins=False)

taperwgi = gf.components.taper(
    length=tapper_l,
    width2=wg_io,
    width1=w2,
    cross_section=cross_section,
    add_pins=False)
taperwgo = gf.components.taper(
    length=tapper_l,
    width2=wg_io,
    width1=w2,
    cross_section=cross_section,
    add_pins=False)

inwgb_r = wgb << bndwg
aux_cs = partial(gf.cross_section.cross_section, width=w2,
cross_section=cross_section)
obndwg = gf.components.bend_s((sbend_l, sbend_h), cross_section=aux_cs
, add_pins=False)
outwgb_r = wgb << obndwg
itapperb_r = wgb << taperwgi
otapperb_r = wgb << taperwgo

inwgb_r.connect("o2", halfp_ref[0].ports["o1"])
outwgb_r.mirror_y()
outwgb_r.connect("o1", halfp_ref[-1].ports["o2"])

inwgb_r_w1.ymin = -w1 / 2
outwgb_r_w1.ymin = -w1 / 2
inwgb_r.y = inwgb_r.y + profile2
outwgb_r.y = outwgb_r.y - profile2

itappert_r.connect("o2", inwgb_r_w1.ports["o2"])
otappert_r.connect("o2", outwgb_r_w1.ports["o2"])

wgt.add_port("o1", port=itappert_r.ports["o1"])
wgt.add_port("o2", port=otappert_r.ports["o1"])

itapperb_r.connect("o1", inwgb_r.ports["o1"])
otapperb_r.connect("o1", outwgb_r.ports["o2"])
```

```
    wgb.add_port("o1", port=itapperb_r.ports["o2"])
    wgb.add_port("o2", port=otapperb_r.ports["o2"])

    # Add wavegudies to contraDC component
    wgt_r = c << wgt
    wgb_r = c << wgb

    # Add optical ports to cdc component
    c.add_port("in", port=wgt_r.ports["o2"])
    c.add_port("drop", port=wgb_r.ports["o2"])
    c.add_port("through", port=wgt_r.ports["o1"])
    c.add_port("add", port=wgb_r.ports["o1"])
    return c.flatten()

if __name__ == "__main__":
    c = simple_contra_dc()
    c.show(show_ports=True, show_subports=True)
```

Author details

Inês Venâncio[1,2,3*†], Joana Tátá[2†], João Santos[2,3], Paulo Duarte[2], Carla Rodrigues[2] and António Teixeira[1,2]

1 IT - Instituto de Telecomunicações, Aveiro, Portugal

2 PICadvanced S.A., Ílhavo, Portugal

3 University of Aveiro, Aveiro, Portugal

*Address all correspondence to: inesvenancio@ua.pt

† These authors contributed equally.

References

[1] Chrostowski L, Hochberg ME. Silicon Photonics Design. Cambridge: Cambridge University Press; 2015. ISBN: 978-1-107-08545-9

[2] Eissa M, Mitarai T, Amemiya T, Miyamoto Y, Nishiyama N. Fabrication of Si photonic waveguides by electron beam lithography using improved proximity effect correction. Japanese Journal of Applied Physics. 2020;**59**(12):126502. DOI: 10.35848/1347-4065/abc78d

[3] Zheng Y, Gao P, Tang X, Liu J, Duan J. Effects of electron beam lithography process parameters on structure of silicon optical waveguide based on SOI. Journal of Central South University. 2022;**29**(10): 3335-3345. DOI: 10.1007/s11771-022-5152-0

[4] Lin S. Computational lithography for silicon photonics design [thesis]. Vancouver: The University of British Columbia; 2020

[5] Sun Y. Introduction to ebeam Lithography. 2018. Available from: https://mrsec.wisc.edu/wp-content/uploads/sites/282/2018/04/Yong-Sun.pdf [Accessed: March 6, 2024]

[6] Totzeck M, Ulrich W, Göhnermeier A, Kaiser W. Pushing deep ultraviolet lithography to its limits. Nature Photonics. 2007;**1**(11):629-631. DOI: 10.1038/nphoton.2007.218

[7] Mack CA. Field Guide to Optical Lithography. Bellingham, Washington: SPIE; 2006. ISBN: 978-0-8194-7821-4. DOI: 10.1117/3.665802

[8] Mack CA. Corner rounding and line-end shortening in optical lithography. In: Microlithographic Techniques in Integrated Circuit Fabrication II. Vol. 4226. Singapore: SPIE; 2006. pp. 83-92. DOI: 10.1117/12.404843

[9] Mistry A, Hammood M, Lin S, Chrostowski L, Jaeger NAF. Effect of lithography on SOI, grating-based devices for sensor and telecommunications applications. In: IEEE 10th Annual Information Technology, Electronics and Mobile Communication Conference (IEMCON). Vancouver, BC, Canada: IEEE; 2019. pp. 932-938. DOI: 10.1109/IEMCON.2019.8936189

[10] Wong AK. Resolution Enhancement Techniques in Optical Lithography. Bellingham, Washington: SPIE; 2001. ISBN: 978-0-8194-7881-8. DOI: 10.1117/3.401208

[11] Simard AD, Ayotte N, Painchaud Y, Bedard S, LaRochelle S. Impact of sidewall roughness on integrated Bragg gratings. Journal of Lightwave Technology. 2011;**29**(24):3693-3704. DOI: 10.1109/JLT.2011.2173556

[12] Pathak S. Chapter 7—Photonics integrated circuits. In: Kaushik BK, editor. Nanoelectronics. San Jose, CA, United States: Elsevier; 2019. ISBN: 978-0-12-813353-8. DOI: 10.1016/B978-0-12-813353-8.00008-7

[13] Saber MG. Silicon photonic devices and circuits for optical communication applications [thesis]. Montréal, Canada: McGill University; 2019

[14] Shi W, Yun H, et al. Ultra-compact, flat-top demultiplexer using anti-reflection contra-directional couplers for CWDM networks on silicon. Optics Express. 2013;**21**(6):6733-6738. DOI: 10.1364/OE.21.006733

[15] Shi W, Wang X, et al. Silicon photonic grating-assisted,

contra-directional couplers. Optics Express. 2013;**21**:3633-3650. DOI: 10.1364/OE.21.003633

[16] Hammood M, Mistry A, et al. Compact, silicon-on-insulator, series-cascaded, contradirectional-coupling-based filters with >50 dB adjacent channel isolation. Optics Letters. 2019; **44**(2):439. DOI: 10.1364/OL.44.000439

[17] Ansys Lumerical MODE. Optical Waveguide & Coupler Solver. Available from: https://www.ansys.com/products/optics/mode [Accessed: March 4, 2024]

[18] GDSfactory Available from: https://gdsfactory.github.io/gdsfactory/index.html [Accessed: March 4, 2024]

[19] Tyona MD. A theoretical study on spin coating technique. Advances in Materials Research. 2013;**2**:195-208. DOI: 10.12989/amr.2013.2.4.195

[20] Levinson HJ. High-NA EUV lithography: Current status and outlook for the future. Japanese Journal of Applied Physics. 2024;**61**:1347-4065. DOI: 10.35848/1347-4065/ac49fa

[21] Hu S, Stricker A, McLean K, Ma C, Roy SG, Percy D, et al. Fully automated in-line optical test system. Advanced materials & photonics. In: 2018 29th Annual SEMI Advanced Semiconductor Manufacturing Conference (ASMC). Saratoga Springs, NY, USA: IEEE; 2018. pp. 399-402. DOI: 10.1109/ASMC.2018.8373215

[22] Dusan G, Xu D, Liboiron-Ladouceur O, Grinberg Y. Deep learning-based prediction of fabrication-process-induced structural variations in nanophotonic devices. ACS Photonics. 2022;**9**:2623-2633. DOI: 10.1021/acsphotonics.1c01973

[23] Peng Y, Pan DZ. ELIAS: An accurate and extensible lithography aerial image simulator with improved numerical algorithms. IEEE Transactions on Semiconductor Manufacturing. 2009;**22**: 276-289. DOI: 10.1109/TSM.2009.2017652

[24] Cauchon J. Silicon photonic Bragg-based devices: Hardware and software [thesis]. Québec: Laval University; 2021

[25] Wei S, Veerasubramanian V, Plant DV, Jaeger NAF, Chrostowski L. Silicon Photonic Bragg-Grating Couplers for Optical Communications. Cambridge, United Kingdom: SPIE OPTO; 2014. DOI: 10.1117/12.2044096